Lecture Notes in Computer Science 15197

Founding Editors

Gerhard Goos
Juris Hartmanis

W0234434

The series Lecture Notes in Computer Science (LNCS), including its subseries Lecture Notes in Artificial Intelligence (LNAI) and Lecture Notes in Bioinformatics (LNBI), has established itself as a medium for the publication of new developments in computer science and information technology research, teaching, and education.

LNCS enjoys close cooperation with the computer science R & D community, the series counts many renowned academics among its volume editors and paper authors, and collaborates with prestigious societies. Its mission is to serve this international community by providing an invaluable service, mainly focused on the publication of conference and workshop proceedings and postproceedings. LNCS commenced publication in 1973.

Federica Proietto Salanitri · Serestina Viriri ·
Ulaş Bağcı · Pallavi Tiwari · Boqing Gong ·
Concetto Spampinato · Simone Palazzo ·
Giovanni Bellitto · Nancy Zlatintsi ·
Panagiotis Filntisis · Cecilia S. Lee ·
Aaron Y. Lee
Editors

Artificial Intelligence in Pancreatic Disease Detection and Diagnosis, and Personalized Incremental Learning in Medicine

First International Workshop, AIPAD 2024
and First International Workshop, PILM 2024
Held in Conjunction with MICCAI 2024
Marrakesh, Morocco, October 10, 2024
Proceedings

 Springer

Editors
Federica Proietto Salanitri ⓘ
University of Catania
Catania, Italy

Ulaş Bağcı ⓘ
Northwestern University
Chicago, IL, USA

Boqing Gong
Boston University
Boston, MA, USA

Google DeepMind
London, UK

Simone Palazzo ⓘ
University of Catania
Catania, Italy

Nancy Zlatintsi ⓘ
National Technical University of Athens
Zografou, Greece

Cecilia S. Lee ⓘ
University of Washington
Seattle, WA, USA

Serestina Viriri ⓘ
University of KwaZulu-Natal
Durban, South Africa

Pallavi Tiwari ⓘ
University of Wisconsin-Madison
Madison, WI, USA

Concetto Spampinato ⓘ
University of Catania
Catania, Italy

Giovanni Bellitto ⓘ
University of Catania
Catania, Italy

Panagiotis Filntisis ⓘ
National Technical University of Athens
Zografou, Greece

Aaron Y. Lee ⓘ
University of Washington
Seattle, WA, USA

ISSN 0302-9743   ISSN 1611-3349 (electronic)
Lecture Notes in Computer Science
ISBN 978-3-031-73482-3   ISBN 978-3-031-73483-0 (eBook)
https://doi.org/10.1007/978-3-031-73483-0

Preface AIPAD 2024

The first Workshop on Artificial Intelligence in Pancreatic Disease Detection and Diagnosis (AIPAD 2024) was held on October 10, 2024 in Marrakech, Morocco, in conjunction with the 27th International Conference on Medical Image Computing and Computer-Assisted Intervention (MICCAI 2024).

In recent years, the field of medical imaging has been revolutionized by the integration of artificial intelligence (AI) and deep learning technologies. These advances have provided healthcare professionals with new tools to improve diagnostic processes and treatment planning. Despite its small size, the pancreas plays a crucial role in both the digestive and endocrine systems. The pancreas is an organ that is distinguished by a high degree of anatomical and functional complexity, which presents a significant challenge in the detection and diagnosis of its diseases, particularly through imaging. Pancreatic disorders, including diabetes, pancreatic cancer, and pancreatitis, represent a significant global health burden, underscored by the considerable diagnostic challenges and high mortality rates associated with these conditions. The advent of deep learning in medical imaging has opened up new avenues for addressing these challenges, offering sophisticated tools with the potential to enhance diagnosis and treatment outcomes for pancreatic diseases.

AIPAD 2024 was organized in response to the increasing need for specific attention to AI applications in pancreatic health. The workshop aimed to serve as a hub for researchers and practitioners to exchange insights, share advances, and collaborate on the frontier challenges associated with pancreatic diseases. We aimed to address the critical issues of organ boundary delineation, motion artifacts, low signal-to-noise ratios, and nuanced integration of multimodal data - all of which are critical in the context of pancreatic disease management.

Following the rigorous standards set by MICCAI, we adopted a double-blind review process to ensure unbiased and fair evaluation, with each submission evaluated by at least three experts in the field. The reviewers, chosen for their proven expertise and significant contributions to medical imaging and AI, were meticulously assigned to avoid any conflicts of interest. The submissions were assessed based on their originality, technical quality, and potential impact on the field. The final decisions regarding acceptance or rejection were made by the Program Chairs based on the rankings, quality, and quantity of submissions.

A total of five accepted papers over six submissions were selected for presentation at the workshop. The accepted papers showcased a diverse array of AI applications, from enhancing detection and segmentation accuracy in complex imaging to optimizing data synthesis for improved model training. These papers collectively represent the workshop's commitment to advancing the precision and efficacy of pancreatic disease diagnosis through cutting-edge AI research.

We would like to express our gratitude to all the authors for their valuable contributions and to the reviewers for their thorough evaluations and constructive feedback. The success of AIPAD 2024 was a result of the collective commitment of the community to advancing pancreatic health through AI. We believe that the proceedings of this workshop will serve as a stimulus for further research and collaboration, ultimately leading to improved outcomes for patients afflicted with pancreatic diseases.

October 2024

Federica Proietto Salanitri
Ulaş Bağcı
Boqing Gong
Concetto Spampinato
Pallavi Tiwari
Serestina Viriri

Preface PILM 2024

The first edition of the Workshop on Personalized Incremental Learning in Medicine (PILM 2024) was held on October 10, 2024, in Marrakesh, Morocco, in conjunction with the 27th International Conference on Medical Image Computing and Computer-Assisted Intervention (MICCAI 2024).

Machine learning models for personalized medicine aim to customize diagnostic and therapeutic strategies to individual patient characteristics. This often requires training models on high-dimensional, sparse, and heterogeneous data. Common approaches involve creating a joint dataset from multiple patients to train a global model, which can then be tailored to individual patients. However, compiling comprehensive datasets for robust machine learning model training is challenging due to data collection and governance issues. Additionally, historical data often shows domain shifts in data distribution due to variations in machinery and diagnostic techniques over time.

Incremental and continual learning paradigms in machine learning have gained significant attention for their potential to develop models that adapt to new information over time without forgetting previously learned knowledge. This capability is well-suited to personalized medicine, where machine learning models could be incrementally trained on data from few or even single patients at a time. Such an approach would protect individual privacy and data ownership while enabling continual model updates as new data becomes available.

During the first edition of PILM workshop, there are 3 accepted papers over 4 submissions. Each paper was presented in-person in Marrakesh by at least one of the co-authors. We followed the same review process as the main MICCAI Conference, using a double-blind procedure with a minimum of three reviews per submission. Reviewers were selected from a pool of researchers with proven expertise in the field and numerous publications on the topics covered by the workshop. Reviewers for each submission were carefully chosen to avoid any potential conflicts of interest. Submissions were ranked based on the received ratings. Final decisions on acceptance or rejection where made by Program Chairs based on the rankings, quality, and number of submissions.

We would like to conclude by thanking the authors for their efforts in producing their contributions and the reviewers for their dedication, patience, and constructive feedback.

October 2024

Giovanni Bellitto
Panagiotis Filntisis
Aaron Y. Lee
Cecilia S. Lee
Simone Palazzo
Nancy Zlatintsi

Organization

Organization Committee AIPAD 2024

Ulaş Bağcı	Northwestern University, USA
Boqing Gong	Google Research, USA
Federica Proietto Salanitri	University of Catania, Italy
Concetto Spampinato	University of Catania, Italy
Pallavi Tiwari	University of Wisconsin-Madison, USA
Serestina Viriri	University of KwaZulu-Natal, South Africa

Program Committee AIPAD 2024

Giovanni Bellitto	University of Catania, Italy
Salvatore Calcagno	University of Catania, Italy
Tolga Çukur	Bilkent University, Turkey
Rutger Hendrix	Università Campus Bio-Medico Rome, Italy
Davood Karimi	Boston Children's Hospital, USA
Aggelos Katsaggelos	Northwestern University, USA
Raffaele Mineo	Università Campus Bio-Medico Rome, Italy
Simone Palazzo	University of Catania, Italy
Matteo Pennisi	Università Campus Bio-Medico Rome, Italy
Ziyue Xu	NVIDIA, USA
Jing Yuan	Zhejiang University, China

Organization Committee PILM 2024

Giovanni Bellitto	University of Catania, Italy
Panagiotis Filntisis	National Technical University of Athens, Greece
Aaron Y. Lee	University of Washington, USA
Cecilia S. Lee	University of Washington, USA
Simone Palazzo	University of Catania, Italy
Nancy Zlatintsi	National Technical University of Athens, Greece

Program Committee PILM 2024

Theodora Chaspari	University of Colorado Boulder, USA
Kalliopi V. A. Dalakleidi	National Technical University of Athens, Greece
Marcella M. Gomez	University of California Santa Cruz, USA
Camila Gonzalez	Stanford University, USA
Mohammad Jafari	Columbus State University, USA
Marc Masana	Graz University of Technology, Austria
Raffaele Mineo	Università Campus Bio-Medico of Rome, Italy
Ana M. Barragán Montero	Université catholique de Louvain, Belgium
Matteo Pennisi	Università Campus Bio-Medico of Rome, Italy
Concetto Spampinato	University of Catania, Italy

Contents

Artificial Intelligence in Pancreatic Disease Detection and Diagnosis

Assessing the Efficacy of Foundation Models in Pancreas Segmentation

Emanuele Rapisarda, Alessandro Giuseppe Gravagno, Salvatore Calcagno$^{(\boxtimes)}$, and Daniela Giordano

University of Catania, Department of Electrical Electronic and Computer Engineering, Via Santa Sofia, 64, 95123 Catania, Italy
salvatore.calcagno@phd.unict.it

Abstract. Accurate pancreas segmentation is crucial for diagnosing and managing pancreatic diseases, facilitating preoperative planning, and aiding transplantation procedures. Effective segmentation enables the identification and monitoring of conditions such as chronic pancreatitis and diabetes mellitus, which are characterized by changes in pancreatic size and volume. Recent advancements in segmentation technology have leveraged foundation models like SAM and MedSAM, achieving state-of-the-art performance in various domains. In this work, we explore the effectiveness of using these models in the particularly challenging domain of pancreas segmentation. We also propose a simple yet effective method for including 3D information into SAM. Our findings suggest that, despite foundation models have a good general knowledge, they are not well-suited for pancreas segmentation without significant architectural modifications and the inclusion of a good prompt. Moreover, we found that simply including volume information significantly enhances segmentation performance, even without the use of an expert prompt.

Keywords: Pancreas Segmentation · MedSAM · Segmentation · Medical Imaging

1 Introduction

Computed tomography (CT) and Magnetic Resonance Imaging (MRI) play critical roles in diagnosing and managing pancreatic diseases. Accurate segmentation of pancreas in these imaging modalities is essential for several reasons. Segmenting the pancreas is crucial for disease diagnosis and monitoring, as identifying and tracking changes in pancreatic volume can indicate conditions like chronic pancreatitis and diabetes mellitus. These diseases often manifest through reductions in pancreatic size and volume [1,2,13]. Preoperative planning also benefits significantly from accurate segmentation, providing detailed anatomical information that aids in the success and safety of surgical procedures like pancreatectomy (partial or total removal of the pancreas). Surgeons need precise data on the pancreas's volume and its relationship with surrounding structures to predict surgical outcomes and minimize complications. Additionally, in pancreas or

F. Proietto Salanitri et al. (Eds.): PILM 2024/AIPAD 2024, LNCS 15197, pp. 3–13, 2025.
https://doi.org/10.1007/978-3-031-73483-0_1

islet cell transplantation, segmentation helps assess the size and suitability of the donor pancreas and monitor post-transplant recovery and tissue integration.

The primary imaging modalities for pancreatic segmentation are MRI and CT scans. The latter are widely available, provide rapid imaging, and are effective in detecting a range of pancreatic diseases, including neoplastic lesions. On the other hand, MRI provides more detailed information compared to CT due to its superior soft tissue contrast. Moreover MRI avoids the risks associated with ionizing radiation exposure inherent in CT scans, making it a safer option, especially for repeated imaging required in chronic disease monitoring.

In recent decades, AI and computer vision algorithms have proven highly effective in segmentation tasks across various scenarios. In the context of pancreatic analysis, these technologies can significantly shorten the time for accurately segmenting regions and detecting abnormalities. However, segmenting the pancreas poses several challenges. First, the pancreas has a complex and variable shape, which can be difficult to delineate accurately. Additionally, MRI images can contain artifacts and noise that vary from slice to slice, complicating the segmentation process. To address this limitation, it is crucial to include volumetric information. By using data from neighboring slices we can confirm or correct ambiguous areas, leading to more accurate segmentation.

Secondly, collecting datasets for medical imaging, particularly for pancreatic abnormalities can be expensive and challenging in terms of time and privacy, limiting the availability of large datasets necessary for effective training. The advent of foundation models have shown to be a promising solution. These models, pre-trained on vast and diverse datasets, can be fine-tuned for specific tasks, thereby recycling general knowledge and reducing the need for extensive domain-specific data. The Segment Anything Model (SAM) [6] exemplifies this approach. However, SAM's effectiveness diminishes when faced with significant domain shifts, such as those encountered in medical imaging. To address this, the MedSAM model was proposed [9], aiming to adapt SAM's capabilities to the medical domain. Despite its advancements, MedSAM currently lacks specific information on pancreatic segmentation, necessitating further adaptation and fine-tuning to ensure accurate and reliable performance in this critical area.

In this work, we explore the potential of directly using foundation models for pancreas segmentation. Specifically, we aim to demonstrate whether the general encoded features remain effective for this specific domain. We first explore the use of expert prompts, either directly or with minimal fine-tuning, as well as an automatic segmentation approach. For the latter, we propose two simple yet effective methods for incorporating volumetric information, thus enhancing segmentation performance.

2   Related Work

Pancreas Segmentation. Recent advancements in segmentation algorithms have significantly enhanced the performance of CT and MRI segmentation tasks [16]. In recent years, pancreas segmentation has primarily been addressed using two principal approaches: convolutional neural networks (CNNs) and transformers. Numerous studies have leveraged CNNs as feature extractors, tipically using hierarchical feature representations to reconstruct segmentation masks [5,12,17]. While some approaches employ 2D U-Net architectures [8,10], 3D-aware models such as 3D U-Net [3] and PankNet [12] have generally achieved superior results. Also, incorporating self-attention layers into these models enhances their ability to capture long-range correlations across the volume, which is crucial for accurately segmenting complex structures like the pancreas. Self-attention mechanisms, as demonstrated in the Vision Transformer (ViT) [4], allow the model to focus on relevant regions across the entire volume, improving segmentation consistency and accuracy. Oktay et al. [11] showcased the effectiveness of self-attention mechanisms in medical imaging with the introduction of the Attention U-Net, which enhances the model's focus on relevant regions, thereby improving segmentation performance. Similarly, Zhou et al. [17] combined 3D U-Net with self-attention layers to achieve state-of-the-art performance in liver and tumor segmentation from CT scans, illustrating the potential of this approach for pancreatic segmentation as well.

Segmentation Foundation Models. The current trend in medical image segmentation is increasingly leaning towards leveraging foundation models, which are pre-trained on large and diverse datasets and can be fine-tuned for specific tasks. Notable among these are the Segment Anything Model (SAM) [6] and its medical variant, MedSAM [9]. These models bring the advantage of generalizability and robustness across different types of segmentation tasks. SAM [6] was trained on natural images, and requires the user to insert a prompt in the form of text, points or bounding boxes. MedSAM [9] extends this capability to the medical domain, offering enhanced performance for medical image segmentation tasks by incorporating domain-specific knowledge and techniques. However, SAM and MedSAM face two primary challenges: they are primarily designed for 2D data and require prompt-based inputs for segmentation. SAM-Med3D [14] and MedLSAM [7], were developed to address the first limitation, but both of them still need a prompt. MedLSAM [7] attempts to create a fully automated segmentation pipeline by initially localizing a region of interest using a bounding box. However, it remains unclear if foundational models are adequate for fully automatic pancreas segmentation. Furthermore, to the best of our knowledge, neither SAM nor MedSAM and their variants have been applied to pancreas segmentation specifically. This work aims to determine whether these methods are still suitable for this task.

3   Method

In this work, we explore the possibility of leveraging Segment Anything Model (SAM) and its specialized version MedSAM to perform pancreas segmentation. Both of them share the same architecture, including an image encoder, a prompt encoder, and a mask decoder. The image encoder processes a 2D slice to generate an internal representation. The prompt encoder encodes the region of interest. The mask decoder then uses the outputs from both the image encoder and the prompt encoder to produce a 2D segmentation mask. We assess their effectiveness and limitations by exploring three main settings: 1) zero-shot usage 2) decoder finetuning and 3) integration of volumetric information.

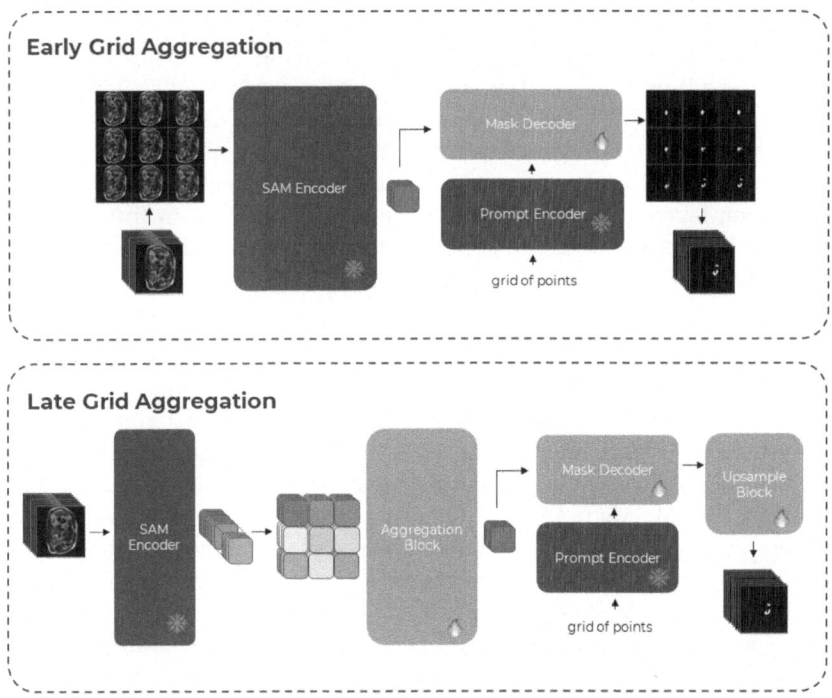

Fig. 1. Overview of Employed Methods for Including Volumetric Information in SAM. In the early fusion strategy, slices are arranged in a 3×3 grid. At the output layer, predictions are cut and rescaled to the original input size. For the late fusion strategy, features are fused after the SAM Encoder using an Aggregation Block. The SAM Decoder then produces a coarse-grained version of the masks, and a final upsampling block restores the depth information.

3.1   Zero-Shot and Finetuning

SAM and MedSAM operate mainly as an assisted segmentation tool for natural and medical images, respectively, requiring input from a domain expert. The expert provides prompts that guide the model to correctly segment the region of interest (ROI), significantly accelerating the segmentation process compared to manual segmentation by the expert alone. However, an important and desirable capability is performing segmentation autonomously, without the need for expert prompts. To evaluate the performance of these foundation models in both scenarios, we crafted two types of prompts: bounding boxes focused on regions of interest (simulating expert guidance) and a fixed prompt (a grid of points evenly spaced across the image) for autonomous segmentation. We tested the models using these prompts in both zero-shot mode and after fine-tuning the decoder.

3.2 Addressing Volumetric Information

The SAM model cannot directly segment an entire MRI sequence as a volume; instead, it segments slice by slice, totally losing the volumetric reference. To address this issue, we implemented two strategies: early grid aggregation and late grid aggregation. An overview of employed methodologies is shown in Fig. 1.

Early Grid Aggregation. For this approach, N adjacent slices are arranged into a single grid image. At this point, the SAM model can be used without any change. During inference, a sliding window approach feeds the network with uniformly sized groups of slices from a single sequence, which are then concatenated into a square grid and fed into the model. By adjusting the window stride, the same slice can appear in multiple groups, providing different predictions based on the context in which it is evaluated. This redundancy allows for the aggregation of multiple predictions, leading to a refined and more accurate final segmentation. This method attempts to capture the relationships between the slices while maintaining the model's architecture. However, since the input resolution is fixed, rescaling operations may result in some loss of information. We chose $N = 9$, thus producing a 3×3 grid.

Late Grid Aggregation. The second approach involves feeding the SAM encoder with one slice at a time. The outputs of N consecutive slices from the SAM encoder are then arranged into a grid and processed by a downscale block to fit the embedding of a single image. This downscale block consists of two basic convolutional layers. The first one applies a 3×3 kernel with dilation matching the input embedding size, thus capturing the relationships between the same areas in different slices. An additional layer is then used for further refinement. The output of this aggregation block goes inside the decoder, which outputs a compact mask for all the slices. An upsample convolutional block attempts to recover the masks for all slices starting from this compact version.

4 Experiments

4.1 Dataset

For our experiments, we used the PanSegData public dataset [15]. It consists of 767 scans from 499 subjects, featuring T1-weighted and T2-weighted abdominal MRI series from five centers, collected between March 2004 and November 2022.

For this work, we considered only the T2W images. The preprocessing pipeline includes bilinear interpolation to uniform the spacing between slices, orientation adjustments to the RAS reference system, and resizing to fit the model input size of 1024×1024. These scans cover a broad region of the human body, not just the pancreas. Typically, the beginning and ending parts of the sequences consist of slices where the pancreas is not present. To address data imbalance, during training we sample either a single slice or a batch of consecutive slices, ensuring positive samples are present at least 85% of the time. Random Flip and Random Rotation are applied with a probability of 50% during training as data augmentation techniques.

4.2   Training Procedure

Except in the case of zero-shot evaluation, we train all employed architectures for 100 epochs using SGD with a learning rate of 0.001 and a weight decay of 0.0005. We use a scheduler that reduces the learning rate by a factor of 10 if the target metric (validation dice score) does not improve for 20 epochs. For all architectures, we aimed to test the generalization capabilities of the SAM Encoder, so we kept both the SAM image and prompt encoders frozen. In the early grid aggregation setting, we trained only the decoder parameters, whereas in the late grid aggregation setting, we also trained the two additional modules. The training objective is the weighted sum of Dice and Cross Entropy Loss. Given that the magnitudes of both losses were similar, we assigned equal weights to them. For the dice loss, we aggregate masks and predictions across the batch size to avoid nonsensical values due to empty predictions or empty masks. The best model is selected based on the highest dice score on the validation set. All training is performed on a NVIDIA RTX A6000. Both SAM and MedSAM require an input prompt. We tested the models in two scenarios: automatic segmentation (without the need for a prompt) and aided segmentation. In the automatic segmentation setting, we provided the prompt encoder with a grid of equally spaced points. In the aided segmentation setting, we used bounding boxes derived from the ground truth, with an added random variation of 25 pixels to simulate expert user inputs.

4.3   Inference and Metrics

During training, for each scan, we sample either a single slice or 9 consecutive slices (when volumetric information is considered). At inference time, we use a sliding window strategy to obtain the final prediction. When processing volumes, we apply a 25% overlap between consecutive windows and average predictions for the same slice. All models were evaluated using Dice score, Jaccard index, precision, and recall.

4.4   Results

Zero-shot Evaluation. We first evaluated SAM and MedSAM in a zero-shot setting, meaning without additional training on pancreas data. While SAM was trained on natural images, MedSAM was trained on medical images, but neither was specifically trained on pancreas datasets.

Both models rely on prior information provided by a prompt. Table 1 showcases the obtained results. When using a generic prompt, such as a grid of points, the output was a segmentation mask that segments any possible object, resulting in poor performance with dice scores close to zero. On the other hand, using bounding boxes improved the results, but despite providing information about the region of interest, the models still struggled to accurately segment the pancreas. This difficulty is attributed to the challenging nature of pancreas segmentation, which is often similar to adjacent tissues.

Decoder Finetuning. We then focused on improving segmentation performance by fine-tuning the decoder parameters while keeping both the SAM Encoder and Prompt Encoder frozen. This approach allowed us to adapt the decoder to better handle pancreas-specific features and nuances without altering the pre-trained encoder components. As indicated in Tab. 1, learning the decoder consistently improves performance across all tested methods and settings. Models utilizing bounding boxes benefit the most from this approach, achieving higher dice scores and jaccard indices compared to those using grid points. This demonstrates the critical role of decoder fine-tuning in optimizing segmentation results.

Table 1. Segmentation results using different prompt types: a grid of points simulates the absence of an expert prompt, while bounding boxes simulate its presence. We evaluate SAM and MedSAM, both in frozen states and with a learnable decoder.

Method	Dice	Jaccard	Precision	Recall
Grid of Points				
SAM	35.78	28.53	49.75	37.01
↪Learn. decoder	60.78	55.90	58.62	67.53
MedSAM	35.78	28.53	49.75	37.01
↪Learn. decoder	59.32	54.79	56.75	69.27
Bounding Boxes				
SAM	78.36	70.32	75.17	83.39
↪Learn. decoder	86.27	78.71	85.19	88.04
MedSAM	78.32	70.30	75.17	83.28
↪Learn. decoder	87.25	79.88	86.40	88.43

Addition of Volumetric Information. Finally, we assessed how incorporating volumetric information during training, without any expert prompts, impacts segmentation performance. First, with early grid aggregation we attempted a straightforward approach (early grid aggregation) by arranging 9 consecutive slices in a 3×3 grid and segmenting the entire grid. The final predictions were then sliced and rescaled to obtain the segmented results. In contrast, late grid aggregation involved fusing features after the encoder by adding an aggregation and an upsampling module to handle volumetric information more effectively. Results with both approaches are shown in Table. 2. We notice that early aggregation is more effective with SAM weights, while MedSAM obtain a marginal gain. This difference may be attributed to their pretraining domains: SAM, which is trained with general knowledge, handles the input grid effectively, whereas Med-SAM, which was trained with individual slices, benefits more from late aggregation. In both cases using late aggregation further improve final performance, thus indicating the effectiveness of our method. Despite these improvements, the final results remain insufficiently accurate. However, it's important to note that expert prompts were not used, and the network had to simultaneously learn both localization and segmentation tasks. This explains the gap in performance with methods in which bounding boxes were used. In Fig. 2, we present some examples of segmentations. The results demonstrate that the progressive addition of modules consistently enhances segmentation quality. Additionally, in some cases, our method achieves performance comparable to or exceeding that of using bounding boxes as prompts. However, qualitatively, our predictions do not differ significantly from those obtained with bounding boxes as the initial prompt.

Table 2. Effect of incorporating volumetric information without an expert prompt: Aggregating consecutive slices either at the input level or in the latent space, significantly enhances segmentation performance. We report segmentation metrics using SAM and MedSAM weights.

Method	Dice	Jaccard	Precision	Recall
SAM	35.78	28.53	49.75	37.01
↪Learn. decoder	60.78	55.90	58.62	67.53
↪Early grid agg.	64.37	58.33	61.58	71.20
↪Late grid agg.	**70.03**	**63.21**	**69.17**	**77.15**
MedSAM	35.78	28.53	49.75	37.01
↪Learn. decoder	59.32	54.79	56.75	69.27
↪Early grid agg.	59.93	55.31	58.26	64.88
↪Late grid agg.	**68.31**	**61.27**	**64.69**	**78.15**

5   Discussion and Conclusion

The evaluation of foundation models such as SAM and MedSAM demonstrates that they are not yet fully ready for zero-shot inference in pancreas segmentation. Despite their strong performance in general image analysis, these models struggle with the specific challenges of pancreas segmentation, particularly when no expert prompts are provided. Incorporating volumetric information during training has shown promise in improving performance, indicating that spatial context plays a crucial role in enhancing segmentation accuracy. However, even with these improvements, the results still fall short of the required accuracy for reliable pancreas segmentation. For future work, several strategies could be explored to address these limitations. One promising direction is to develop learned prompts tailored for automatic segmentation, which could enhance the

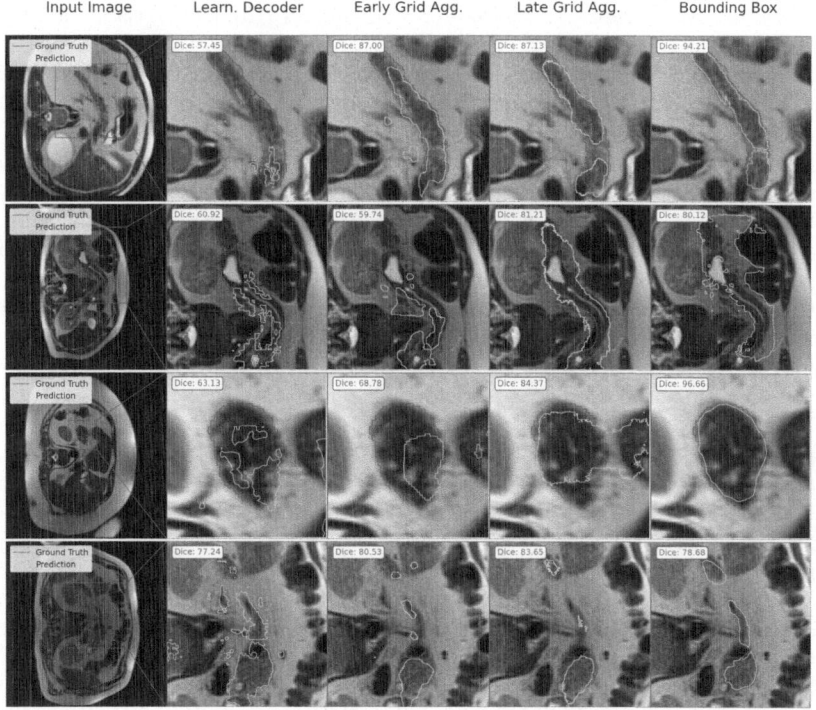

Fig. 2. Qualitative results of segmentation. The first two rows use the SAM Encoder, while the last two rows use the MedSAM Encoder. In columns we present in order, the input image and predictions from different methods: finetuning only the decoder, early grid aggregation, late grid aggregation, and using a bounding box as a prompt. The first and third rows illustrate cases where our late aggregation method yields better results than using a bounding box, whereas the second and fourth rows show the opposite. In all cases, including volume information significantly enhances segmentation quality, even without an expert prompt.

model's ability to handle specific segmentation tasks without expert intervention. Additionally, a two-stage approach - first for region of interest computation and then for detailed segmentation - could offer more precise results. These approaches could bridge the current gaps and advance the field towards more effective and accurate zero-shot segmentation models.

Acknowledgments. S. Calcagno acknowledges financial support from PNRR MUR project PE0000013-FAIR.

References

1. Busireddy, K.K., et al.: Pancreatitis-imaging approach. World J. Gastrointest. Pathophysiol. **5**(3), 252–270 (2014)
2. Campbell-Thompson, M.L.: The influence of type 1 diabetes on pancreatic weight. Diabetologia **59**(1), 217–221 (2016)
3. Çiçek, Ö., Abdulkadir, A., Lienkamp, S.S., Brox, T., Ronneberger, O.: 3D U-Net: learning dense volumetric segmentation from sparse annotation. In: Ourselin, S., Joskowicz, L., Sabuncu, M.R., Unal, G., Wells, W. (eds.) MICCAI 2016. LNCS, vol. 9901, pp. 424–432. Springer, Cham (2016). https://doi.org/10.1007/978-3-319-46723-8_49
4. Dosovitskiy, A., et al.: An image is worth 16x16 words: transformers for image recognition at scale. http://arxiv.org/abs/2010.11929
5. Farag, A., Lu, L., Roth, H.R., Liu, J., Turkbey, E., Summers, R.M.: A bottom-up approach for pancreas segmentation using cascaded superpixels and (Deep) image patch labeling. IEEE Trans. Image Process. **26**(1), 386–399 (2017)
6. Kirillov, A., et al.: Segment Anything. In: 2023 IEEE/CVF International Conference on Computer Vision (ICCV), pp. 3992–4003 (2023)
7. Lei, W., Wei, X., Zhang, X., Li, K., Zhang, S.: Medlsam: Localize and segment anything model for 3d medical images. arXiv preprint arXiv: (2023)
8. Lim, S.H., Kim, Y.J., Park, Y.H., Kim, D., Kim, K.G., Lee, D.H.: Automated pancreas segmentation and volumetry using deep neural network on computed tomography. Sci. Rep. **12**(1), 4075 (2022)
9. Ma, J., He, Y., Li, F., et al.: Segment anything in medical images. Nat. Commun. **15**, 654 (2024)
10. Nishio, M., Noguchi, S., Fujimoto, K.: Automatic pancreas segmentation using coarse-scaled 2D model of deep learning: usefulness of data augmentation and deep u-net. Appl. Sci. **10**(10) (2020)
11. Oktay, O., et al.: Attention U-Net: learning where to look for the pancreas. In: MIDL (2018)
12. Proietto Salanitri, F., Bellitto, G., Irmakci, I., Palazzo, S., Bagci, U., Spampinato, C.: Hierarchical 3D feature learning for pancreas segmentation. In: Lian, C., Cao, X., Rekik, I., Xu, X., Yan, P. (eds.) MLMI 2021. LNCS, vol. 12966, pp. 238–247. Springer, Cham (2021). https://doi.org/10.1007/978-3-030-87589-3_25
13. Sasamori, H., Fukui, T., Hayashi, T., Yamamoto, T., Ohara, M., Yamamoto, S., Kobayashi, T., Hirano, T.: Analysis of pancreatic volume in acute-onset, slowly-progressive and fulminant type 1 diabetes in a Japanese population. J. Diabetes Investig. **9**(5), 1091–1099 (2018)
14. Wang, H., et al.: Sam-med3D (2023)

15. Zhang, Z., et al.: Large-scale multi-center CT and MRI segmentation of pancreas with deep learning (2024)
16. Zhang, Z., Yao, L., Keles, E., Velichko, Y., Bagci, U.: Deep learning algorithms for pancreas segmentation from radiology scans: a review. Adv. Clin. Radiol. **5**(1), 31–52 (2023)
17. Zhou, Y., et al.: Hyper-pairing network for multi-phase pancreatic ductal adenocarcinoma segmentation. Lecture Notes in Computer Science (including subseries Lecture Notes in Artificial Intelligence and Lecture Notes in Bioinformatics) **11765 LNCS**, pp. 155–163 (2019). https://doi.org/10.1007/978-3-030-32245-8_18

Hybrid Deep Learning Model for Pancreatic Cancer Image Segmentation

Wilson Bakasa[ID], Clopas Kwenda[ID], and Serestina Viriri[✉][ID]

School of Mathematics, Statistics and Computer Science, University of KwaZulu-Natal, Durban, South Africa
{219098448,221072651}@stu.ukzn.ac.za, viriris@ukzn.ac.za

Abstract. Pancreatic cancer remains one of the most challenging malignancies to diagnose and treat, necessitating advances in medical imaging techniques for early and accurate detection. This study presents a novel hybrid approach to pancreatic cancer histopathology image segmentation by integrating deep neural networks with traditional machine learning models. Our method leverages the strengths of both paradigms to enhance segmentation performance. Specifically, we employ supervised learning to train deep convolutional neural networks (CNNs), namely ResNet50 and VGG16, to extract high-level feature vectors from medical histopathology images obtained from The Cancer Imaging Archive pancreatic-ct dataset by the National Institutes of Health Clinical Center. These feature vectors serve as inputs to various machine learning classifiers, including Random Forest, K-Nearest Neighbors (KNN), XGBoost, Linear Support Vector Machine (LinearSVM), Linear Discriminant Analysis (LDA), and Gaussian Naive Bayes (GNB). Combining the feature extraction capabilities of deep learning models with the decision-making prowess of traditional classifiers, our hybrid framework based on XGBoost produced the best segmentation results among other classifiers by achieving a precision value of (0.96), F1-Score value of (0.97) and a recall value of (0.98). Extensive experiments and cross-validation on benchmark datasets demonstrate that our approach outperforms standalone models, showcasing its potential in clinical applications for improved diagnostic accuracy (0.936) and patient outcomes.

Keywords: Segmentation · ResNet50 · VGG16 · Feature Extraction · XGBoost · Histopathology Images · Pancreatic Cancer

1 Introduction

In recent years, advancements in medical imaging and artificial intelligence (AI) have shown promising results in assisting medical professionals with the early detection and accurate segmentation of cancerous tissues. Among the various

F. Proietto Salanitri et al. (Eds.): PILM 2024/AIPAD 2024, LNCS 15197, pp. 14–24, 2025.
https://doi.org/10.1007/978-3-031-73483-0_2

types of cancer, pancreatic cancer remains particularly challenging due to its subtle early symptoms and the difficulty in obtaining accurate imaging-based diagnostics [3, 7]. Integrating deep neural networks (DNNs) with traditional machine learning (ML) models presents a compelling approach to enhance the accuracy and efficiency of pancreatic cancer image segmentation [9] This paper explores the feasibility and performance of hybrid DNN-ML models for pancreatic cancer image segmentation, contributing to the evolving field of medical AI by bridging the gap between advanced deep learning techniques and established ML methodologies in the context of oncological imaging.

2   Related Work

In the histopathology of PDAC following neoadjuvant therapy (NAT) [8], this study examines whether artificial intelligence-based residual tumour burden segmentation could provide a more objective and repeatable TRS technique. Modified U-nets with a DenseNet161 encoder produced the highest mean segmentation accuracy results. An overall multiclass average F1 score of 0.82 was obtained for the tumour tissue segmentation, with a high mean F1 score of 0.86. An algorithm known as the hybrid Deep Convolutional Neural Network with Deep Belief Network (DCNN_DBN) [20] is used to detect pancreatic cancer. Based on trial results, the current CAD system achieves 99.6% accuracy and presents enormous potential and safety in the automated detection of malignant and benign tumours. This classifier greatly reduces the computational complexity. More anomalies in pancreatic cancer cells could be found by improving the recommended method. The research [6] presents the first deep convolutional neural network design to categorise and segment pancreatic histopathology pictures using a sizable WSI dataset. 100% classification accuracy was attained by the WSI-level technique and 95.3% by the automatic patch-level approach. Furthermore, to ascertain which regions of an image are more crucial for PDAC detection, they represented the segmentation and classification results of histological images. Experimental findings demonstrate the suggested model's ability to accurately detect PDAC using histopathological pictures, highlighting the promise of this useful application.

Researchers have used convolutional neural networks (CNNs) like YOLOv3 and modified AlexNet to classify pancreatic ductal adenocarcinoma from CT scans with high accuracy and sensitivity [4]. A self-supervised learning algorithm called "pseudo-lesion segmentation" was developed to address the need for large annotated datasets. This boosted the performance of both CNN and transformer-based deep learning models for pancreatic cancer classification on internal and external validation datasets [21]. A study proposes a hybrid model combining Attention U-Net and TAU-Net for segmenting PDAC mass and surrounding vessels in CT images. The model uses a 3D-CNN aggregator network to combine the outputs of the two segmentation networks, achieving a Dice score of 86.93% for PDAC mass segmentation [16]. A collaborative framework is presented that combines meta-heuristic optimization algorithms with deep learning models for accurate segmentation and classification of breast cancers. The framework uses a hybrid optimization strategy to optimize the performance of the deep learning models [10]. Pyramidal format histopathology is imitated by integrating multiscale methods and deformable convolution to improve classification accuracy. This method, validated by real clinical datasets, offers a robust and generalizable tool for pancreatic cancer diagnosis with an accuracy of up to 96% [11].

3   Methods and Techniques

The experiment was conducted on the Google Colab platform, which provides free TPU and GPU resources in the cloud. Given the high computational demands of the experiment, NVIDIA Tesla GPU of 16 Gigabytes RAM acceleration was utilized [15]. The backend platform for TensorFlow GPU and Keras GPU was used as the deep learning API. A standard, publicly available dataset was sourced from The Cancer Imaging Archive pancreatic-ct dataset by the National Institutes of Health Clinical Center [2]. Feature extractors such as VGG16 and ResNet-50 models were used to identify and isolate important characteristics or attributes from raw data, transforming it into a format that the machine learning classifier can analyze and use to make predictions for the segmentation task.This study chose these two pre-existing models due to their innately dissimilar architecture that abstracts unrelated information from images used for object detection purposes [13]. XGBoost algorithm was adopted in the proposed system to perform the segmentation task. Studies carried out by [14] conducted a study to compare the efficacy and effectiveness machine learning classifiers for object detection demonstrated that XGBoost lagorithm had efficacy level that above 90%, while the other algorithms had efficacy levels below 90%. It is against this backdrop that the proposed model has advocated towards XGBoost.

3.1   Feature Extractors

VGG16 and ResNet50 deep neural networks adopted under transfer learning were used to extract a set of features for the classifiers to perform the segmentation task. These models were chosen because they have an innately dissimilar architecture that abstracts unrelated information from the images used for object detection purposes [12].

3.2   XGBoost

Aims to minimize a regularized objective function that balances model complexity and prediction accuracy [22]. The objective function $\mathcal{L}$ for a given iteration t can be expressed as:

$$\mathcal{L}^t = \sum_{i=1}^{n} \ell(y_i, \hat{y}^{(t-1)} + f_t(x_i) + \Omega(f_t)) \tag{1}$$

Where: n is the number of observations. ℓ is a differentiable convex loss function that measures the difference between the prediction $\hat{y}^{(t-1)} + f_t(x_i)$ and the actual target y_i. $\hat{y}^{(t-1)}$ is the prediction of the $i-th$ observation at the previous iteration. $f_t(x_i)$ is the prediction from the $t-th$ tree. $\Omega(f_t$ is the regularization term for the complexity of the tree f_t.

This robust framework allows XGBoost to deliver high accuracy while maintaining computational efficiency [1], making it a popular choice for many machine-learning tasks.

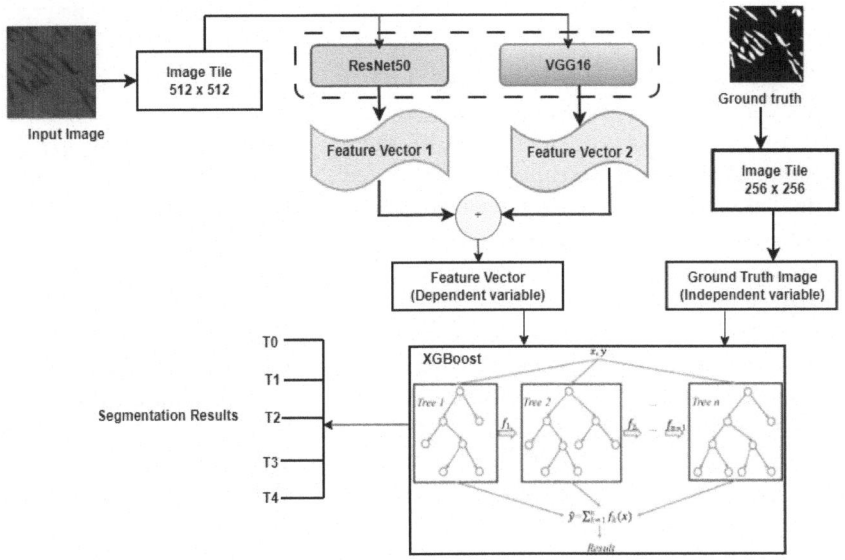

Fig. 1. The workflow of the proposed model

3.3   Proposed Hybrid Model

Features extracted from the images using ResNet-50 and VGG16 were fused and inputted to XGBoost to segment the pancreas histopathology images.

The proposed model is shown in Fig. 1. Several tests were conducted to determine the performance of the proposed model. The dataset was split into 80% for training purposes and 20% for testing purposes. Other experiments were performed by extracting features using trained models from ResNet50 and VGG16. The model identified five regions, each represented by distinct pixel values: 0, 64, 128, 191, and 255. Regions with a pixel value of 0 (black) contain healthy cells, while regions with a pixel value of 255 (white) indicate full-blown pancreatic cancer. The pixel values (0, 64, 128, 191, and 255) were converted to class labels 0, 1, 2, 3, and 4, respectively as shown in Table 1.

Table 1. Class label description

Class Label	Description
0	Healthy cells
1	Cells are 25 % progressive to cancer
2	Cells are 50 % progressive to cancer
3	Cells are 75 % progressive to cancer
4	Cells have full-blown cancer

4   Experimental Results and Discussion

Due to the varying sizes of the images, all the images were resized to 512×512 to maintain universal uniformity. ResNet50 and VGG16 models adopted under transfer learning were employed to generate a set of features for machine learning classifiers to perform segmentation on pancreas cancer images. VGG 16 model produced a set of 128 features while the ResNet50 model generated a set of 64 features. Features obtained from these models were concatenated together to produce a final feature vector of 192. The final feature vector was then utilized by machine learning classifiers such as XGBoost (Extreme Gradient Boosting), RF (Random Forest), LDA (Linear Discrimination Analysis), GNB (Gaussian Naive Bayes), LinearSVM (Linear Support Vector Machine), and KNN (k-Nearest Neighbor) to perform the segmentation task. Metrics such as Accuracy, RMSE (Root Mean Square Error), Precision, Recall, and F1-Score were used to measure the segmentation performance of each machine learning classifier. Precision is a metric used to evaluate the accuracy of the model. It is the ratio of correctly predicted positive observations to the total predicted positives. Tables 2, 3, 4, 5, 6, 7 show the segmentation performance of RF, KNN, LinearSVM, LDA, GNB, and XGBoost based models in terms of Precision, Recall, and F1 -score concerning five regions of interest. Categorising is into five different tumours (T), node (N), and metastases (M) (TNM) staging system class labels [19], which are T0 (category 0), T1 (category 1), T2 (category 2), T3 (category 3), and T4 (category 4).

All six models managed to detect regions in Category 0 and Category 4, with GNB achieving the best performance for Category 0 by attaining a precision of 0.98, followed by LDA (0.97), XGBoost (0,96), RF (0.96), KNN (0.96) and LinearSVM (0.90). For Category 4, the LinearSVM-based model recorded the best precision value of 0.87, followed by RF (0.77), KNN (0.77), XGBoost 0.77, LDA (0.76), and GNB (0.68). Concerning recall, linearSVM based model produced the best recall value of 1.00 for category 0 followed by RF (0.98), KNN (0.98), XGBoost (0.98), LDA (0.96), and GNB (0.89). For Category 4, XGBoost performed well as it obtained a recall value of 0.81, followed by RF (0.80), GNB (0.75), LDA (0.71), and GNB (0.41). F1-Score is closely linked to accuracy; table results show that the XGBoost-based model achieved the best results for both categories by attaining an F1-score of 0.97 for T0 and 0.79 for T4. RF model closely followed by achieving an F1-score of 0.97 for T0 and 0.78 for T4.

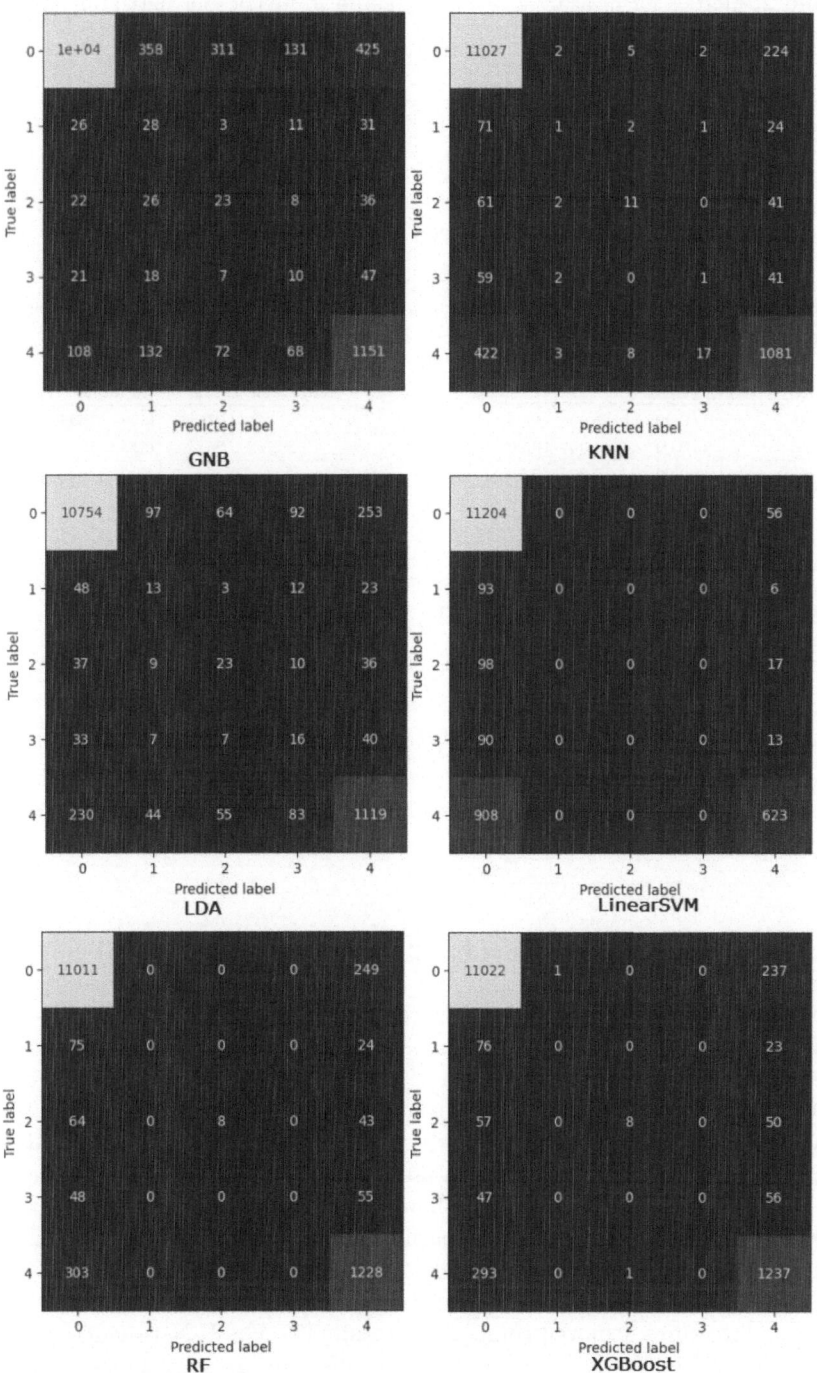

Fig. 2. Confusion Matrix for RF, XGBoost, LDA, GNB, LinearSVM, and KNN

Table 2. Metrics for measuring segmentation through RF

Pixel Region	Precision	Recall	F1-Score
0	0.96	0.98	0.97
1	0.00	0.00	0.00
2	1.00	0.07	0.13
3	0.00	0.00	0.00
4	0.77	0.80	0.78

Table 3. Metrics for measuring segmentation through KNN

Pixel Region	Precision	Recall	F1-Score
0	0.96	0.98	0.96
1	0.10	0.01	0.02
2	0.42	0.10	0.16
3	0.05	0.01	0.02
4	0.77	0.71	0.71

Table 4. Metrics for measuring segmentation through LinearSVM

Pixel Region	Precision	Recall	F1-Score
0	0.90	1.00	0.95
1	0.00	0.00	0.00
2	0.00	0.00	0.00
3	0.00	0.00	0.00
4	0.87	0.41	0.55

Table 5. Metrics for measuring segmentation through LDA

Pixel Region	Precision	Recall	F1-Score
0	0.97	0.96	0.96
1	0.08	0.13	0.13
2	0.15	0.20	0.17
3	0.08	0.16	0.18
4	0.76	0.73	0.75

Table 6. Metrics for measuring segmentation through GNB

Pixel Region	Precision	Recall	F1-Score
0	0.98	0.89	0.93
1	0.05	0.28	0.08
2	0.06	0.02	0.09
3	0.04	0.10	0.06
4	0.68	0.75	0.71

Table 7. Metrics for measuring segmentation through XGBoost

Pixel Region	Precision	Recall	F1-Score
0	0.96	0.98	0.97
1	0.00	0.00	0.00
2	0.89	0.07	0.13
3	0.00	0.00	0.00
4	0.77	0.81	0.79

A set of confusion matrices in Fig. 2 demonstrates that the XGBoost-based model achieved the best performance by correctly classifying most pixels (12267) into their respective categories and closely followed by RF(12247). GNB-based classifier received the worst performance as it misclassified the most number of pixels(1861).

Confusion matrix results go incongruent with the results displayed in Table 8, which indicates the hybridized model of deep neural networks with XGBoost classifier produced the best results in terms of accuracy (94%) and recorded the lowest RMSE value of 0.86. Fig. 3 shows the masks and the predicted images by the five different classifiers. Comparing the results of our hybrid model with related studies reveals the effectiveness of our approach. For instance, a study on breast cancer histopathology image segmentation using U-Net and autoencoder architectures achieved competitive results [5]. The U-Net model, which is widely recognized for its performance in medical image segmentation, showed high precision and recall values similar to those achieved in our hybrid model. In another comparison, deep learning models like ResNet50, ResNet101, VGG16,

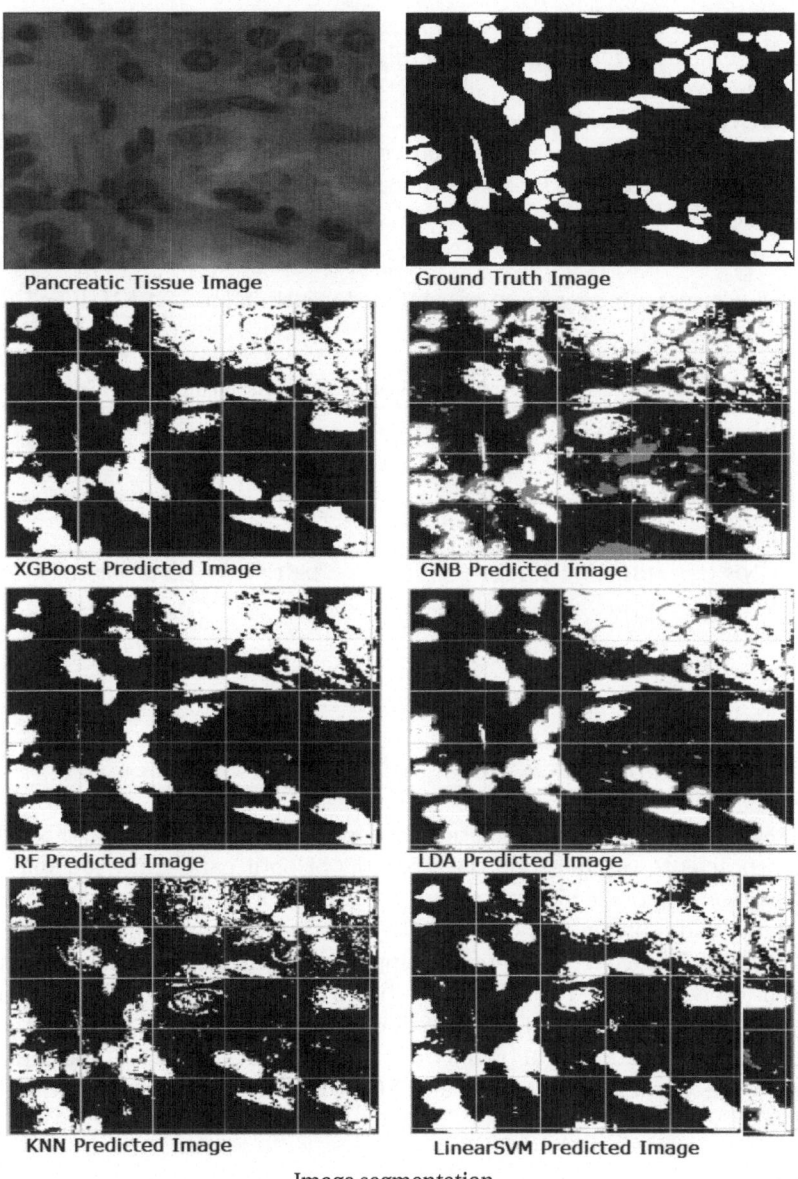

Image segmentation.

Fig. 3. Pancreatic Cancer Histopathology

and VGG19 were used for breast cancer detection using histopathology images. The models were evaluated on metrics such as accuracy, AUC, recall, and loss [17]. Although these models also performed well, they typically require more computational resources and extensive training data compared to the XGBoost framework. Furthermore, a study utilizing a highly reliable random subspace classifier ensemble for breast cancer diagnosis from biopsy images reported notable performance but didn't surpass the recall and F1-Score values achieved by your XGBoost-based framework [18]. Overall, our XGBoost-based approach provides a robust and efficient alternative to deep learning models, maintaining high precision, recall, and F1-Score in histopathology image segmentation while potentially being more resource-efficient Table 8.

Table 8. Comparative assessment of hybridized models in terms of RMSE and Accuracy

Model	Total Pixels	Correctly Classified	Misclassified	RMSE	Accuracy
RF	13108	12247	861	0.876	0.934
KNN	13108	12121	987	0.959	0.923
LinearSVM	13108	11827	1281	1.134	0.902
LDA	13108	11925	1183	1.450	0.910
GNB	13108	11247	1861	1.816	0.858
XGBoost	13108	12267	841	0.860	0.936

5 Conclusion

By integrating deep feature extraction with traditional ML classifiers, this hybrid approach aims to capitalize on the strengths of both paradigms: the ability of DNNs to learn intricate hierarchical representations from raw data and the interpretability and robustness of classical ML algorithms in handling structured feature vectors. The effectiveness of this approach lies in its potential to improve segmentation accuracy, reduce computational complexity, and enhance clinical decision-making in the realm of pancreatic cancer diagnosis and treatment planning. The best results were obtained from fusing features extracted from ResNet-50 and VGG16 and inputting them into XGBoost (which outperformed the other models).

References

1. Chen, T., Guestrin, C.: Xgboost: A scalable tree boosting system. In: Proceedings of the 22nd ACM Sigkdd International Conference on Knowledge Discovery and Data Mining, pp. 785–794 (2016)
2. Clark, K., Vendt, B.: The cancer imaging archive (TCIA): maintaining and operating a public information repository. J. Digital Imaging (2013). https://doi.org/10.1007/s10278-013-9622-7

3. Daher, H., et al.: Advancements in pancreatic cancer detection: integrating biomarkers, imaging technologies, and machine learning for early diagnosis. Cureus **16**(3) (2024)
4. Dinesh, M., Bacanin, N., Askar, S., Abouhawwash, M.: Diagnostic ability of deep learning in detection of pancreatic tumour. Sci. Rep. **13**(1), 9725 (2023)
5. Drioua, W.R., Benamrane, N., Sais, L.: Breast cancer histopathological images segmentation using deep learning. Sensors **23**(17), 7318 (2023)
6. Fu, H., et al.: Automatic pancreatic ductal adenocarcinoma detection in whole slide images using deep convolutional neural networks. Front. Oncol. **11**, 665929 (2021)
7. Hameed, B.S., Krishnan, U.M.: Artificial intelligence-driven diagnosis of pancreatic cancer. Cancers **14**(21), 5382 (2022)
8. Janssen, B.V.: Artificial intelligence-based segmentation of residual tumor in histopathology of pancreatic cancer after neoadjuvant treatment. Cancers **13**(20), 5089 (2021)
9. Jiang, X., Hu, Z., Wang, S., Zhang, Y.: Deep learning for medical image-based cancer diagnosis. Cancers **15**(14), 3608 (2023)
10. Khan, S., et al.: Bilevel hyperparameter optimization and neural architecture search for enhanced breast cancer detection in smart hospitals interconnected with decentralized federated learning environment. IEEE Access (2024)
11. Kou, Y., Xia, C., Jiao, Y., Zhang, D., Ge, R.: Dactransnet: a hybrid CNN-transformer network for histopathological image classification of pancreatic cancer. In: CAAI International Conference on Artificial Intelligence, pp. 422–434. Springer (2023). https://doi.org/10.1007/978-981-99-9119-8_38
12. Kwenda, C., Gwetu, M., Fonou-Dombeu, J.V.: Ontology with deep learning for forest image classification. Appl. Sci. **13**(8), 5060 (2023)
13. Kwenda, C., Gwetu, M., Fonou-Dombeu, J.V.: Hybridizing deep neural networks and machine learning models for aerial satellite forest image segmentation. J. Imaging **10**(6), 132 (2024)
14. Łoś, H., et al.: Evaluation of XGBoost and LGBM performance in tree species classification with sentinel-2 data. In: 2021 IEEE International Geoscience and Remote Sensing Symposium IGARSS, pp. 5803–5806. IEEE (2021)
15. Madiajagan, M., Raj, S.S.: Parallel computing, graphics processing unit (GPU) and new hardware for deep learning in computational intelligence research. In: Deep Learning and Parallel Computing Environment for Bioengineering Systems, pp. 1–15. Elsevier (2019)
16. Mahmoudi, T., et al.: Segmentation of pancreatic ductal adenocarcinoma (PDAC) and surrounding vessels in CT images using deep convolutional neural networks and texture descriptors. Sci. Rep. **12**(1), 3092 (2022)
17. Mahmud, M.I., Mamun, M., Abdelgawad, A.: A deep analysis of transfer learning based breast cancer detection using histopathology images. In: 2023 10th International Conference on Signal Processing and Integrated Networks (SPIN), pp. 198–204. IEEE (2023)
18. Rachapudi, V., Lavanya Devi, G.: Improved convolutional neural network based histopathological image classification. Evol. Intel. **14**(3), 1337–1343 (2021)
19. Rosen RD, S.A.: TNM classification, StatPearls Publishing (2024). https://www.ncbi.nlm.nih.gov/books/NBK553187/
20. Thanya, T., Wilfred Franklin, S.: Novel computer aided diagnostic system using hybrid neural network for early detection of pancreatic cancer. Automatika: časopis za automatiku, mjerenje, elektroniku, računarstvo i komunikacije **64**(4), 815–826 (2023)

21. Viriyasaranon, T., Woo, S.M., Choi, J.H.: Unsupervised visual representation learning based on segmentation of geometric pseudo-shapes for transformer-based medical tasks. IEEE J. Biomed. Health Inform. **27**(4), 2003–2014 (2023)
22. Zhang, P., Jia, Y., Shang, Y.: Research and application of XGBoost in imbalanced data. Int. J. Distrib. Sens. Netw. **18**(6), 15501329221106936 (2022)

Leveraging SAM and Learnable Prompts for Pancreatic MRI Segmentation

Cristian Delle Castelle, Fabio Spampinato, Federica Proietto Salanitri$^{(\boxtimes)}$ (ID),
Giovanni Bellitto(ID), and Concetto Spampinato(ID)

University of Catania, Catania, Italy
`federica.proiettosalanitri@unict.it`

Abstract. Accurate segmentation of the pancreas in magnetic resonance imaging (MRI) is essential for enhancing diagnostic and therapeutic strategies in pancreatic diseases. In this study, we explore the application of the Segment Anything Model (SAM), a state-of-the-art foundation model, for pancreas segmentation in MRI scans. We present a preliminary approach that utilizes AutoSAM, a recent work designed to optimize input prompts for the SAM decoder, aiming to improve segmentation capabilities. To evaluate the performance of our method, we employ a publicly available MRI dataset, allowing for comparison with existing segmentation techniques. Preliminary results suggest that learned prompts may lead to potential improvements in pancreas segmentation, indicating the promise of foundation models in medical imaging tasks.

Keywords: Pancreas Segmentation · Magnetic Resonance Imaging · Foundation Models

1 Introduction

Segmentation of the pancreas in abdominal MRI scans plays a crucial role in the diagnosis and treatment of pancreatic diseases. Traditionally, this task has relied primarily on CT [8, 11, 16, 19] scans due to their high contrast resolution. However, MRI offers a promising alternative with its superior soft tissue contrast, particularly in T1-weighted (T1W) and T2-weighted (T2W) modalities. The ability of MRI to provide more detailed anatomical information and better soft tissue differentiation is critical for accurate pancreatic segmentation, which is essential for identifying lesions, masses, and other anomalies that directly impact the diagnosis and treatment of pancreatic diseases.

In this preliminary study, our objective is to investigate the potential of the SAM [7] (Segment Anything Model) foundation model for the challenging task of pancreas segmentation in MRI scans. SAM has gained recognition for its adaptability in segmenting a diverse range of objects across various contexts.

C. D. Castelle and F. Spampinato—Equal contribution.

© The Author(s), under exclusive license to Springer Nature Switzerland AG 2025
F. Proietto Salanitri et al. (Eds.): PILM 2024/AIPAD 2024, LNCS 15197, pp. 25–34, 2025.
https://doi.org/10.1007/978-3-031-73483-0_3

Given and image and a visual prompt specifying what to segment, SAM employs an image encoder and a prompt encoder to create embeddings, which are then combined in a lightweight mask decoder to predict segmentation masks. The effectiveness of the SAM model stems from its training on a large-scale dataset containing over one billion masks and 11 million natural images. However, in other domains such as medical image analysis, large-scale dataset of this magnitude may not be available, which is why SAM struggles to generalize to medical image segmentation.

Thus, adapting SAM to the domain of medical imaging presents a distinct set of challenges. Medical images frequently exhibit variations in contrast, noise, and artifacts that are not typically present in the datasets on which SAM was originally designed. Consequently, a careful tuning of the model to the specific textures and patterns of medical scans is necessary.

Another challenge associated with SAM is the necessity for prompts, such as bounding boxes, points, text or masks, which guide the model during the segmentation process. Although SAM is capable of performing segmentation with as little as a single point, the creation of these prompts in a medical context still requires time and expertise, often necessitating the involvement of a medical professional. To address this bottleneck, we explore the use of AutoSAM [13], an innovative iteration of the SAM model that attempts to overcome the need for manually created prompts. AutoSAM is designed to learn how to generate its own prompt embeddings from the available data.

The efficacy of our strategy is validated through the use of MRI scans obtained from a publicly available dataset [19]. The performance of AutoSAM is compared with traditional segmentation methods, employing standard metrics such as the Dice coefficient and Jaccard index to provide a quantitative assessment of its effectiveness.

2   Related Work

Foundation Models in Medical Imaging. The introduction of Foundation Models (FMs) has opened new frontiers in the field of artificial intelligence (AI). These models are designed to learn from extensive and diverse datasets through self-supervised learning techniques, offering a more scalable and efficient alternative to traditional deep learning paradigms. However, despite their promising capabilities, the application of foundation models in medical imaging presents unique challenges, primarily due to the differences in data characteristics compared to natural images.

The Segment Anything Model (SAM) [7] represents a key example of a foundation model, initially designed for robust image segmentation tasks across natural image datasets, SAM operates as a promptable segmentation method that requires input points or bounding boxes to define the segmentation targets. While this approach resembles traditional interactive segmentation techniques, SAM demonstrates superior generalization abilities. Although SAM has demonstrated efficacy in numerous segmentation tasks within the domain of natural images, many studies have utilized SAM "out of the box" for various medical image segmentation tasks, revealing contrastive results. For instance, concurrent studies have conducted a comprehensive evaluation of SAM across a diverse range of medical images [4,10,14,20], showing that while SAM achieved satisfactory segmentation results for targets with clear boundaries, it struggled significantly with medical targets characterized by weak boundaries or low contrast.

In order to address the limitations of SAM, a specialized adaptation has been proposed, namely MedSAM [9], designed to enhance segmentation performance across a range of medical imaging tasks, including organ segmentation and lesion detection. MedSAM represents an improvement over SAM, as it has been fine-tuned on an extensive dataset comprising over one million medical image-mask pairs, thereby significantly enhancing its segmentation capabilities. It is worth to note, however, that MedSAM has not yet been specifically applied to pancreas segmentation. This is a particularly challenging task due to the pancreas's complex anatomy, its proximity to other organs, and the inherent low contrast often present in MRI scans. These factors can complicate the accurate delineation of the pancreas, making it a critical area for further research.

The adaptation of SAM for use in medical imaging tasks requires the implementation of additional strategies, such as fine-tuning or prompt engineering, to enhance its performance on specific medical datasets. The investigation of these adaptations represents an active area of research, as evidenced by the work of Zhang [18], who discusses the challenges and perspectives of utilizing foundation models in medical image analysis.

To further address the limitations associated with prompting in SAM, AutoSAM [13] is introduced as an extension to SAM. Its purpose is to improve segmentation performance by generating optimized input prompts specifically for medical imaging tasks. In AutoSAM, the original prompts encoder is replaced with a prompt generator network using the same input image. This approach was evaluated on three medical datasets, including RGB images from histopathology and gastrointestinal imaging. Even though these images are from a completely separate domain, features and patterns they contain may share similarities with those in natural images, potentially facilitating the adaptation of SAM for these specific tasks.

In contrast, our study proposes the application of AutoSAM [13] to MRI scans of the pancreas. This presents a significant challenge, as MRI images differ markedly from natural images and introduce unique difficulties for segmentation. Specifically, pancreatic MRI scans often exhibit noise, poorly defined boundaries, and variations in contrast. Additionally, the position and shape of the pancreas can vary significantly between patients, factors which greatly complicate the accurate delineation of this organ.

Pancreas Segmentation. Pancreas segmentation is a crucial task in medical imaging, particularly for the diagnosis and management of conditions such as pancreatic cancer. Several studies have made significant advancements in this area, employing a range of imaging modalities, including computed tomography

(CT) [6,8,15,16] and magnetic resonance imaging (MRI) [1,2,11,19]. Never-theless, despite the expansion of research in this field, the number of studies dedicated to MRI pancreas segmentation remains relatively limited. One of the primary challenges in pancreas segmentation using MRI is the inherent charac-teristics of the imaging modality. The contrast, noise, and poorly defined bound-aries inherent to MRI scans complicate the accurate delineation of the pancreas from surrounding tissues. Despite these challenges, recent studies have demon-strated the effectiveness of deep learning approaches in improving the accuracy of segmentation. For example, Zhang et al. [19] conducted a large-scale multi-center study focusing on pancreas segmentation using both CT and MRI. Their research employs deep learning techniques to achieve robust segmentation across diverse datasets, addressing the challenges presented by MRI, such as the need for precise boundary identification when dealing with varying tissue contrasts. In a related study, Ji et al. [6] proposed the ResDAC-Net, a novel model that utilizes residual double asymmetric spatial kernels for pancreas segmentation. This innovative approach aims to enhance the precision of pancreas segmenta-tion by effectively capturing the intricate anatomical structures of the pancreas, thereby mitigating the challenges posed by varying image qualities and improving overall accuracy. Furthermore, PankNet [11] proposed a hierarchical 3D feature learning approach that has been shown to improve the accuracy of pancreas delineation. The method captures relevant features across multiple scales, which is particularly beneficial in managing the anatomical variations and complexities often seen in MRI scans. Although the existing literature demonstrates encour-aging developments in pancreas segmentation, particularly through the use of deep learning techniques, continued research is necessary to further refine these methods and address unique challenges associated with MRI.

3   Method

In this study, we used the AutoSAM [13] framework, an adaptation of the Seg-ment Anything Model (SAM) [7], to perform pancreas segmentation in magnetic resonance imaging (MRI). AutoSAM leverages the capabilities of SAM by using a prompt generation network to improve segmentation performance through adap-tive prompt embeddings.

To tailor the AutoSAM architecture for MRI applications, we implemented adaptations to handle single-channel image inputs. Specifically, we replaced the first layers of both the image encoder F_i and the prompt encoder F_p with custom layers designed to process single-channel data. The weights of these layers were initialized by averaging the pre-trained weights of their respective counterparts in the original model. This initialization strategy facilitates effective transfer learning, allowing the model to leverage learned features from natural images while adapting to the medical imaging domain.

The segmentation process begins with the generation of prompt embeddings Z_p and image embedding Z_i from the input MRI image I:

$$Z_p = F_p(I) \qquad (1)$$

$$Z_i = F_i(I) \tag{2}$$

These embeddings are subsequently combined and and fed into the prompt decoder within the SAM framework D, which uses both the image features and the generated prompts to compute the predicted segmentation mask M_{pred}:

$$M_{\text{pred}} = D(Z_i, Z_p) \tag{3}$$

To assess the model's performance, we employed the Dice loss function, which quantifies the overlap between the predicted segmentation mask M_{pred} and the ground truth mask M_{gt}. The Dice loss is expressed mathematically as:

$$L_D = 1 - \frac{2 \cdot |M_{\text{pred}} \cap M_{\text{gt}}| + \epsilon}{|M_{\text{pred}}| + |M_{\text{gt}}| + \epsilon} \tag{4}$$

where $|M|$ represents the area of the respective masks, and ϵ is a small constant introduced for numerical stability. This loss function effectively captures the performance of the segmentation model by emphasizing the importance of correctly identifying the pancreas region.

Our primary objective is to exploit the knowledge embedded in the SAM framework to tackle the specific and complex task of pancreas segmentation without reliance on manually annotated prompts from medical experts. By employing a network that autonomously generates prompts directly from the MRI images, we aim to streamline the segmentation process and reduce the burden of annotation in clinical settings.

4   Experimental Results

4.1   Dataset

The dataset used in this study consists of 767 abdominal MRI scans (385 T1-weighted and 382 T2-weighted) from 499 patients, collected from five different institutions in the United States [19]. The two modalities provide complementary information, making them more (or less) suitable depending on the task at hand. For instance, compared to T1W, T2W scans better delineate the contours of the pancreas from surrounding tissues, making them the ideal choice for developing segmentation models. The dataset includes a wide range of patient anatomies and imaging parameters, reflecting the heterogeneity encountered in clinical practice. The annotations were performed by a team of five radiologist using ITK-SNAP [17] and validated by an additional senior radiologist.

4.2   Training Procedure

Our segmentation framework was trained, using the 382 T2W MRIs, by dividing the dataset into 70% for training, 10% for validation, and 20% for testing. We standardized the orientation of the MRI scans according to the RAS convention

and normalized voxel intensities to fall in the range of 0 to 1. During training, voxel resampling was performed to achieve uniform isotropic (1 mm) voxel
spacing across MRI scans, which also served as a data augmentation strategy,
along with random horizontal flipping and 90-degree rotations, to increase the
robustness of the model to input variation.

Before passing them to the image encoder, MRI images are resized to 1024 ×
1024 pixels to match the input size used by SAM [7], ensuring compatibility and
optimal use of the encoder's learned features. For the prompt generator network,
the input slices are resized to 512 × 512 pixels.

During training, gradients are backpropagated through the decoder to the
prompt generator network, ensuring that only the prompt generator network is
free to learn, while both the SAM's image encoder and mask decoder are kept
frozen throughout the training process. By limiting the learning process to the
generator, we focus the training effort on generating effective prompt vectors
that guide SAM's decoder to produce accurate segmentations.

We used a stochastic gradient descent (SGD) optimizer with a learning rate
of 0.01 and a batch size of 4 to minimize the dice loss over 300 epochs. Model
selection was guided by the best Dice score on the validation dataset.

For each epoch, we randomly sample a single slice from each MRI scan to
compute the loss between the slice and its corresponding ground truth. During
the validation and testing phases, we process each slice of an MRI scan individually, reconstructing the entire volume and calculating a Dice score for the entire
MRI. For quantitative evaluation, we report the Dice score and the Intersec-
tion over Union (IoU) as metrics. All experiments are performed on an NVIDIA
A100 GPU, and the proposed approach is implemented in PyTorch using the
MONAI [3] framework.

4.3 Results

The results of this study show that the proposed model is promising for MRI-
based pancreas segmentation. Table 1 shows that our model achieves performance metrics on par with state-of-the-art models such as U-Net [12] and nnU-
Net [5], and even shows comparable performance to PanSegNet [19], a specialized
3D model designed for this task.

Our model, trained according to our learning strategy, outperforms both zero-
shot SAM [7] and MedSAM [9] models. This is indicative of our model's ability
to adapt to the challenges of domain shift, which is particularly relevant when
applying pre-trained models to medical images, a domain very different from
the data on which they were originally trained. Furthermore, the superior performance of our model compared to MedSAM [9], which was trained on medical
images, suggests that domain-specific training is crucial, especially for challenging organs such as the pancreas with often blurry boundaries. Figure 1 provides
a qualitative assessment of our model's performance, showing its robustness to
a variety of cases, including MRI scans of patients with cysts or tumors within
the pancreas. In the majority of cases presented, the model demonstrates a high

Table 1. Segmentation performance comparison between our approach and other state-of-the-art methods using T2W MRI images. Our method, exhibits competitive performance, against both 2D and 3D methods.

Method	Modality	Dice (%)	Jaccard (%)	Precision (%)	Recall (%)
UNet[12]	2D	59.36	47.20	64.52	71.48
white nnUNet[5]	2D	80.96	81.98	82.01	83.11
SAM w/ bbox	2D	74.62	70.05	75.63	81.85
white MedSAM w/ bbox	2D	74.57	70.13	75.74	81.87
PanSegNet[19]	3D	86.01	86.78	85.77	85.88
white **Our**	2D	81.67	73.53	77.14	89.96

degree of accuracy, confirming its potential effectiveness even in pathological scenarios.

However, the images in the last row of Fig. 1 show cases where the model tends to oversegment, particularly in cases with blurred boundaries. These exam-

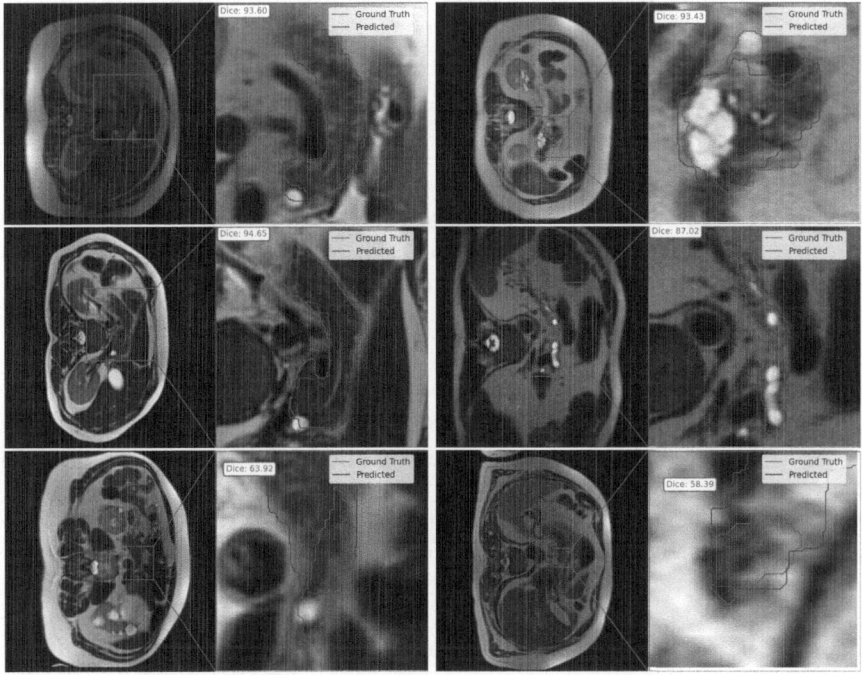

Fig. 1. Qualitative results of pancreas segmentation using our approach on T2-weighted (T2W) MRI images. The last row shows two cases where the model fails, indicating oversegmentation of the pancreas.

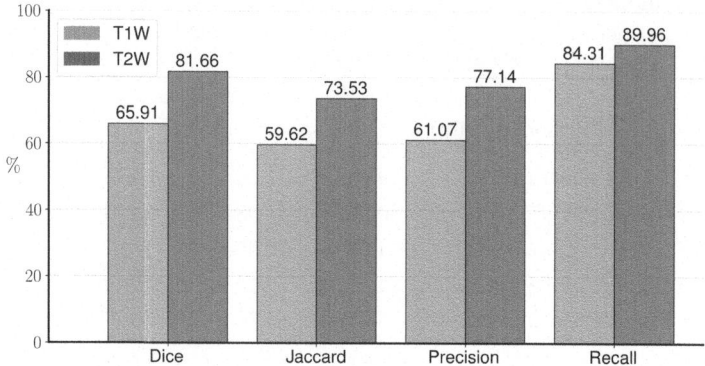

Fig. 2. Impact of MRI Modality on Segmentation: Performance comparison when using T1-weighted (T1W) or T2-weighted (T2W) MRI scans as input, highlighting the enhanced segmentation accuracy with T2W scans.

ples highlight the current limitations of the model and underscore the need for further refinement. A potential strategy to improve the accuracy of the model could be the implementation of targeted preprocessing techniques to improve image contrast, which may help to more accurately identify organ boundaries. In addition, the relatively low precision score suggests that the model may be generating false positives, possibly due to the model's high sensitivity to features that resemble pancreatic tissue. This could be addressed by refining the model's classification threshold or by incorporating post-processing steps to filter out spurious segmentations.

Incorporating volumetric information could also address the issue of over-segmentation, as it would allow the model to leverage spatial continuity and context, potentially reducing false positives and improving the overall segmentation quality.

Ablation Studies. We additionally assessed the impact of image modality on the performance of our segmentation model. Figure 2 presents a comparative analysis between the use of T1-weighted (T1W) and T2-weighted (T2W) MRI scans. The model's performance is consistently superior with T2W images, as evidenced by the Dice Score and Jaccard index, indicating a more accurate representation of the pancreatic tissue.

The enhanced contrast of structures containing water in T2W images probably contributes to this improved performance, allowing for a better differentiation between the pancreas and adjacent tissues. In contrast, the attenuated contrast in T1W images appears to present a challenge to the model's ability to correctly segment the pancreas, resulting in a reduction in precision and recall. These results highlight the crucial role of image modality in automated pancreas segmentation and suggest the potential benefits of integrating multiple

MRI modalities to improve model accuracy. Future work may include the development of multimodal approaches that leverage the specific advantages of each imaging modality to further enhance the reliability and usability of segmentation models in clinical settings.

5   Conclusion

In our work, we applied an existing approach of training a prompt generator to adapt a pre-trained frozen model to the challenging task of pancreas segmentation from MRI scans. With limited data available, this method effectively exploits the knowledge of large-scale pre-trained models, avoiding extensive retraining, while still catering to our specialized application.

The results confirm that our adapted model achieves competitive performance, demonstrating the potential of using foundation models in data-constrained medical imaging scenarios. Although we encountered over-segmentation in regions with unclear boundaries, our approach provides a promising direction for future research.

Going forward, we will focus on refining prompt generation to increase model specificity and exploring the broader applicability of this technique to other complex medical imaging tasks, with the goal of improving diagnostic tools and patient outcomes.

Acknowledgement. F. Proietto Salanitri, G. Bellitto and C. Spampinato acknowledge financial support from PNRR MUR project PE0000013-FAIR.

References

1. Asaturyan, H., Gligorievski, A., Villarini, B.: Morphological and multi-level geometrical descriptor analysis in CT and MRI volumes for automatic pancreas segmentation. Comput. Med. Imaging Graph. **75**, 1–13 (2019)
2. Cai, J., Lu, L., Xie, Y., Xing, F., Yang, L.: Improving deep pancreas segmentation in CT and MRI images via recurrent neural contextual learning and direct loss function. arXiv preprint arXiv:1707.04912 (2017)
3. Cardoso, M.J., et al.: Monai: An open-source framework for deep learning in healthcare. arXiv preprint arXiv:2211.02701 (2022)
4. Deng, R., et al.: Segment anything model (SAM) for digital pathology: assess zero-shot segmentation on whole slide imaging. arXiv preprint arXiv:2304.04155 (2023)
5. Isensee, F., Jäger, P.F., Full, P.M., Vollmuth, P., Maier-Hein, K.H.: nnU-net for brain tumor segmentation. In: Brainlesion: Glioma, Multiple Sclerosis, Stroke and Traumatic Brain Injuries: 6th International Workshop, BrainLes 2020, Held in Conjunction with MICCAI 2020, Lima, Peru, October 4, 2020, Revised Selected Papers, Part II 6, pp. 118–132. Springer (2021). https://doi.org/10.1007/978-3-030-72087-2_11
6. Ji, Z., et al.: Resdac-net: a novel pancreas segmentation model utilizing residual double asymmetric spatial kernels. Med. Bio. Eng. Comput. 1–14 (2024)
7. Kirillov, A., et al.: Segment anything. In: Proceedings of the IEEE/CVF International Conference on Computer Vision. pp. 4015–4026 (2023)

8. Liu, Z., et al.: Pancreas segmentation in CT based on RC-3Dunet with SOM. Multimedia Syst. **30**(2), 66 (2024)
9. Ma, J., He, Y., Li, F., Han, L., You, C., Wang, B.: Segment anything in medical images. Nat. Commun. **15**(1), 654 (2024)
10. Mazurowski, M.A., Dong, H., Gu, H., Yang, J., Konz, N., Zhang, Y.: Segment anything model for medical image analysis: an experimental study. Med. Image Anal. **89**, 102918 (2023)
11. Proietto Salanitri, F., Bellitto, G., Irmakci, I., Palazzo, S., Bagci, U., Spampinato, C.: Hierarchical 3d feature learning forpancreas segmentation. In: Machine Learning in Medical Imaging: 12th International Workshop, MLMI 2021, Held in Conjunction with MICCAI 2021, Strasbourg, France, September 27, 2021, Proceedings 12, pp. 238–247. Springer (2021). https://doi.org/10.1007/978-3-030-87589-3_25
12. Ronneberger, O., Fischer, P., Brox, T.: U-net: convolutional networks for biomedical image segmentation. In: Medical image computing and computer-assisted intervention–MICCAI 2015: 18th international conference, Munich, Germany, October 5-9, 2015, proceedings, part III 18, pp. 234–241. Springer (2015). https://doi.org/10.1007/978-3-319-24574-4_28
13. Shaharabany, T., Dahan, A., Giryes, R., Wolf, L.: Autosam: Adapting sam to medical images by overloading the prompt encoder. arXiv preprint arXiv:2306.06370 (2023)
14. Wald, T., et al.: Sam. md: Zero-shot medical image segmentation capabilities of the segment anything model. In: Medical Imaging with Deep Learning, Short Paper Track (2023)
15. Wang, Y., et al.: Pancreas segmentation using a dual-input v-mesh network. Med. Image Anal. **69**, 101958 (2021)
16. Yang, E., et al.: NNU-net-based pancreas segmentation and volume measurement on CT imaging in patients with pancreatic cancer. Acad. Radiol. (2024)
17. Yushkevich, P.A., Gao, Y., Gerig, G.: ITK-snap: an interactive tool for semi-automatic segmentation of multi-modality biomedical images. In: 2016 38th Annual International Conference of the IEEE Engineering in Medicine and Biology Society (EMBC), pp. 3342–3345. IEEE (2016)
18. Zhang, S., Metaxas, D.: On the challenges and perspectives of foundation models for medical image analysis. Med. Image Anal. 102996 (2023)
19. Zhang, Z., et al.: Large-scale multi-center ct and mri segmentation of pancreas with deep learning. arXiv preprint arXiv:2405.12367 (2024)
20. Zhou, T., Zhang, Y., Zhou, Y., Wu, Y., Gong, C.: Can sam segment polyps? arXiv preprint arXiv:2304.07583 (2023)

Optimizing Synthetic Data for Enhanced Pancreatic Tumor Segmentation

Linkai Peng[1], Zheyuan Zhang[1], Gorkem Durak[1], Frank H. Miller[1],
Alpay Medetalibeyoglu[1], Michael B. Wallace[2], and Ulas Bagci[1(✉)]

[1] Department of Radiology, Northwestern University, Chicago, IL, USA
ulas.bagci@northwestern.edu

[2] Division of Gastroenterology and Hematology, Mayo Clinic, Jacksonville, FL, USA

Abstract. Pancreatic cancer remains one of the leading causes of cancer-related mortality worldwide. Precise segmentation of pancreatic tumors from medical images is a bottleneck for effective clinical decision-making. However, achieving a high accuracy is often limited by the small size and availability of real patient data for training deep learning models. Recent approaches have employed synthetic data generation to augment training datasets. While promising, these methods may not yet meet the performance benchmarks required for real-world clinical use. This study critically evaluates the limitations of existing generative-AI based frameworks for pancreatic tumor segmentation. We conduct a series of experiments to investigate the impact of synthetic *tumor size* and *boundary definition* precision on model performance. Our findings demonstrate that: (1) strategically selecting a combination of synthetic tumor sizes is crucial for optimal segmentation outcomes, and (2) generating synthetic tumors with precise boundaries significantly improves model accuracy. These insights highlight the importance of utilizing refined synthetic data augmentation for enhancing the clinical utility of segmentation models in pancreatic cancer decision making including diagnosis, prognosis, and treatment plans. Our code will be available at https://github.com/lkpengcs/SynTumorAnalyzer.

Keywords: Medical image segmentation · Tumor synthesis · Diffusion model · Generative AI · Pancreas tumors

1 Introduction

Pancreatic cancer, the most dangerous tumor type, leads to merely a 12% five-year survival rate regardless of the stage, according to the report from the American Cancer Society's 2023 Cancer Facts [11,23]. Particularly, the survival rate drops to 3% in distant (stage IV or metastatic) pancreatic tumors. Thus, it is vital for clinicians to detect pancreatic cancer at early stages. Computed Tomography (CT) imaging methods serve as the fundamental tool for early detection

L. Peng and Z. Zhang—Contributed equally. This work is supported by the NIH funding: R01-CA246704, R01-CA240639, U01-DK127384-02S1, and U01-CA268808.

F. Proietto Salanitri et al. (Eds.): PILM 2024/AIPAD 2024, LNCS 15197, pp. 35–44, 2025.
https://doi.org/10.1007/978-3-031-73483-0_4

due to their non-invasive and low-cost properties [27]. However, the low contrast in medical images creates a unique challenge in finding tumor regions in CT scans. Effective pancreas tumor segmentation is crucial for treatment planning, as it provides essential clinical information, including precise tumor volumes. Since the deep learning age emerged, many automatic pancreas tumor segmentation methods have been proposed [7, 10, 19, 25, 27]. Medical segmentation datasets, including CT and MRI modalities, are also publicly available for the pancreas [1, 20, 26]. However, pancreas scans with tumor cases are still limited due to data privacy concerns [1]. Thus, synthetic tumor for pancreatic cancer attracts significant attention.

Several deep learning-based approaches have been proposed for synthesizing diseased regions in various organs. For instance, Shen et al. and Lyu et al. focused on lung tumor generation [16, 22], Shang et al. addressed the fundus [21], Billot et al. proposed methods for brain lesion synthesis brain [2]. Similarly, liver and other organs are studied by [9, 12, 17]. Focusing on pancreatic tumors, Lai et al. [13] introduced a rule-based approach that simulates tumor growth, invasion, and death. The method assigns states to each pixel within the pancreas and evolves them based on predefined rules, generating tumors at various stages. Li et al. [15] took a different approach. Authors first select locations within the pancreas and generate textures, subsequently refining shapes through morphological operations. Statistical analysis during shape generation allows for better control and more realistic rendering of both tumor size and shape. Chen et al. [4] employed an autoencoder model combined with a latent diffusion model [18] to synthesize realistic tumors with similar location selection process and shape generation by [9]. Wu et al. [24] further advanced this method by replacing the diffusion model with a generative adversarial network (GAN) [6]. The proposed method incorporated adversarial training through a pre-trained segmentor to enhance the reliability of the synthesized tumors.

What do we propose? Leveraging the current tumor generation tools, we carefully investigate the limitations inherent in generation-based frameworks and present a series of experiments designed to test specific hypotheses. In particular, we investigate the impacts of *synthetic tumor size variation* and the *precision of tumor boundary definitions* on tumor segmentation accuracy. Our findings indicate that (1) selecting an optimal combination of tumor sizes is crucial for achieving superior segmentation outcomes, and (2) precise tumor boundary annotations significantly enhance model performance.

2   Methods

Problem Definition. We explore potential reasons for the unsatisfactory performance of pancreatic tumor segmentation models with tumor synthesis under real-world scenarios. Based on this analysis, we carefully evaluate the influence of tumor size and the non-perfect tumor annotations.

Influence of Tumor Size. The size of synthetic tumors significantly influences the training of segmentation models. This factor becomes more critical when dealing with pancreatic tumors. These tumors are relatively small compared to

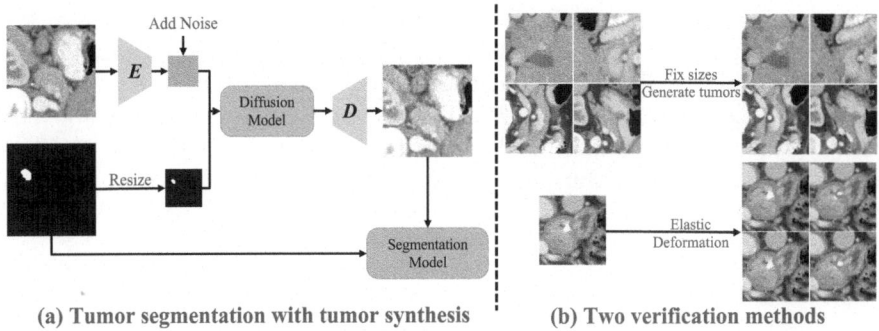

(a) Tumor segmentation with tumor synthesis   (b) Two verification methods

Fig. 1. Schematic demonstration of our proposed verification strategy. Panel **(a)** shows the tumor segmentation process using a diffusion model to synthesize pancreatic tumors. E and D denote the encoder and decoder of a pre-trained autoencoder. Panel **(b)** depicts two proposed verification methods. The upper part shows the generation of fixed-size tumors for segmentation. The lower part illustrates the elastic deformation used for generating noisy labels.

the entire pancreas, which presents a considerable challenge for segmentation. From our observations, current methods all generate the tumor regions first and then generate the tumor textures. Given that the synthesis of the tumor texture also relies on the initial generation of the tumor region(s), the small size of these pancreatic tumors further compounds the complexity of the segmentation process.

Influence of Tumor Annotations. Defining the boundary of synthetic tumors is another critical factor that influences segmentation models. The precision of boundary definitions directly affects a model's ability to delineate tumor margins accurately, essential for clinical applications. Since current models solely focus on the regions within the generated tumor masks, the boundaries may suffer from abrupt changes and unrealistic appearances, leading to inferior segmentation performances.

  Hypotheses: Based on our observations, we propose two hypotheses as follows:

- **Hypothesis 1**: Effectively utilizing different synthetic tumor sizes can help improve the segmentation performances.
- **Hypothesis 2**: More accurate tumor boundaries can lead to better performances.

2.1   Tumor Generation Network

In order to carefully investigate the influence of tumor size and annotation precision, we follow the widely used framework in *Difftumor* [4] for tumor generation. An input volume x is initially processed through the encoder E of a pre-trained autoencoder, which compresses it into latent features x_l within the latent space. Simultaneously, a corresponding tumor mask y is generated and resized to align with the dimensions of x_l. These two components are integrated to form the masked features $z = (1 - y) \otimes x_l$. Subsequently, z is fed into a pre-trained latent diffusion model [18]. The resultant output is then upsampled through the autoencoder's decoder D to produce the final synthetic volume $\hat{x}$. $\hat{x}$ and y are then used to train a segmentation model, which is supervised using both Dice loss and Cross Entropy loss. Our proposed verification methods overview is shown in Fig. 1.

2.2   Dataset

Following the methodology outlined in [4], we utilize the MSD-Pancreas [1] dataset for real tumor data, while employing samples from the Pancreas-CT [20] and BTCV [14] datasets as healthy controls. The MSD-Pancreas dataset comprises of 282 volumetric (3D) CT scans with publicly available pixel-level annotations for pancreas and tumor regions. In contrast, the Pancreas-CT dataset includes 82 abdominal contrast-enhanced volumetric CT scans provided by the NIH team, and the BTCV dataset contains 30 volumetric CT scans with manually annotated abdominal organs including pancreas.

2.3   Impact of Tumor Size

In the original tumor generation pipeline, four pre-defined radii—r_{tiny}, r_{small}, r_{medium}, and r_{large}—are specified. One of these radii is randomly chosen as $r_{selected}$. Subsequently, the radii along the x, y, and z axes are generated based on a uniform distribution as shown in Eq. 1

$$r_x, r_y, r_z \sim \text{Uniform}(r_{selected} - \Delta, r_{selected} + \Delta), \tag{1}$$

where Δ defines the variability range around the selected radius $r_{selected}$, and is different for different tumor sizes. With this way, the model can be fed with tumors of various sizes.

To verify our hypothesis 1, we control the tumor size during the generation process. Initially, we analyze tumor volumes from the MSD-Pancreas dataset [1], arranging them in ascending order of volume. Quartile values are then computed to categorize the tumors into four predefined size classes: tiny, small, medium, and large. Associated radii—r_{tiny}, r_{small}, r_{medium}, and r_{large}—are determined for each category. These quartile-based radii replace the initially predefined radii to ensure an accurate representation of natural tumor size variations. This methodology facilitates the controlled synthesis of tumors, thus allowing for a systematic evaluation of the performances of segmentation models across a spectrum of tumor sizes.

2.4   Impact of Boundary

In the original framework, tumor boundaries are refined through elastic deformation, introducing a randomized, non-linear transformation to the synthetic tumor regions. This is achieved by manipulating a grid of control points, where each control point is subjected to random shifts governed by a normal distribution with a specified standard deviation σ. The shifts generate a displacement field $\mathbf{D}$, described mathematically as Eq. 2

$$\mathbf{D}(i, j) = (\Delta x, \Delta y) \sim \mathcal{N}(0, \sigma^2), \qquad (2)$$

where (i, j) are the grid coordinates, and $(\Delta x, \Delta y)$ represents the displacement at each point, drawn from a normal distribution with mean 0 and variance σ^2. This displacement field dictates the relocation of pixels within the original tumor region, thereby creating more realistic boundaries that closely mimic those of actual tumors.

To validate our second hypothesis, we apply the same elastic deformation process to the original tumor labels in the MSD-Pancreas dataset [1], resulting in (artificially) noisy labels. Given that the labels are three-dimensional, we apply elastic deformation to x, y, and z dimensions independently. Notably, we exclusively use real tumor images paired with the modified labels for training, omitting any synthetic volumes.

3   Experiments and Results

3.1   Training Protocol

We partition the MSD-Pancreas [1] dataset into training and testing subsets using a 4:1 ratio. Selected samples from the Pancreas-CT [20] dataset and the BTCV [14] dataset are used for synthesizing data. The intensities of all scans are clipped to the range [-175, 250] and then normalized to [0, 1]. During training, all inputs are randomly cropped to a size of 96 × 96 × 96 with an equal ratio of tumor to normal tissue regions. We also apply data augmentation techniques such as random rotation, random flip and random shift. For the autoencoder model [3], we adopt the Vector Quantized Generative Adversarial Networks (VQGAN) [5] architecture to learn to map the original 3D volumes to a shared latent space. For the diffusion model [8], we use the latent diffusion model [18] structure with inputs from the compressed latent space of the autoencoder. For segmentation models, we use U-Net [19], nnU-Net [10], and SwinUNETR [7]. All methods are trained on an A100 GPU for 200 epochs.

3.2   Influence of Tumor Size

We trained U-Net [19], nnU-Net [10], and SwinUNETR [7] using synthetic tumor volumes of varying sizes: tiny, small, medium, large, and mixed, where the Mixed category aligns with the original *Difftumor* [4] method. Table 1 presents the performance outcomes, highlighting both the Dice Similarity Coefficient (DSC) and Normalized Surface Distance (NSD) metrics. The results indicate a clear improvement in segmentation performance for all models when synthetic volumes are utilized. Notably, models using large synthetic tumor sizes achieved

Table 1. Segmentation performances on various tumor sizes. We report the Dice Similarity Coefficient (DSC) and Normalized Surface Distance (NSD). We can observe a clear improvement in segmentation performance for all models when synthetic volumes are utilized.

Method	Tumor Size	DSC(%)	NSD(%)
Without synthetic volumes			
U-Net	-	48.48	43.26
nnU-Net	-	50.50	47.32
SwinUNETR	-	45.88	39.76
With synthetic volumes			
U-Net	Tiny	47.92	43.62
	Small	50.07	46.45
	Medium	51.32	47.77
	Large	54.84	49.76
	Mixed	54.49	49.41
nnU-Net	Tiny	50.98	46.02
	Small	51.45	46.91
	Medium	53.67	49.59
	Large	55.99	50.89
	Mixed	52.10	47.14
SwinUNETR	Tiny	54.77	49.81
	Small	52.76	49.24
	Medium	55.16	49.29
	Large	56.10	53.58
	Mixed	55.50	51.03

superior outcomes compared to those with mixed-size tumors, suggesting that larger synthetic tumors might be more effective in improving model accuracy. This observation underscores the impact of utilizing varied synthetic tumor sizes on segmentation performance, indicating that the optimal combination of these sizes could be crucial for achieving the best segmentation results.

Qualitative visualizations of the segmentation results from all compared methods are presented in Fig. 2. We provide the tumor segmentation results for U-Net [19], nnU-Net [10], and SwinUNETR [7]. These results demonstrate that the incorporation of synthetic volumes significantly enhances the models' ability to accurately segment tumors, as evidenced by the more precise delineation of tumor boundaries.

3.3   Influence of Noisy Label

We trained U-Net [19], nnU-Net [10], and SwinUNETR [7] using noisy labels generated by elastic deformation. Only CT scans from the MSD-Pancreas [1] are used for training, and we artificially add label noise into the annotation mask. The results, detailed in Table 2, indicate a decline in performance for all models when trained with noisy labels. This can also be confirmed, as shown in Fig. 2, and the segmentation accuracy declines when the models are trained with noisy labels, leading to less precise boundary delineations.

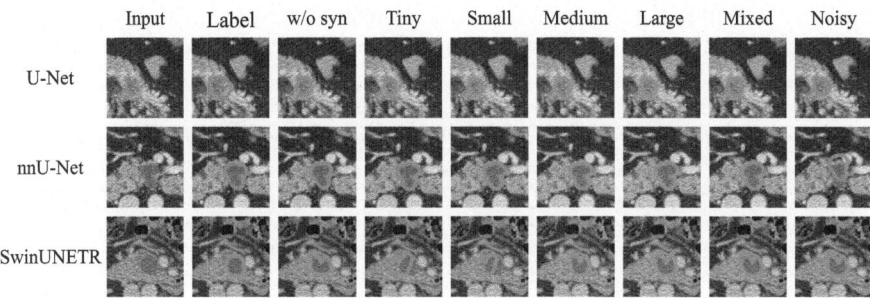

Fig. 2. Qualitative visualization results of all compared methods. The rows represent the models used, while the columns display results from left to right: raw input volumes, labels, results without synthetic volumes, and segmentation for tumors categorized as Tiny, Small, Medium, Large, Mixed, and with Noisy labels. Tumor boundaries are delineated in the figure.

Additionally, we conducted an ablation study to further evaluate these models across various noise levels: low, moderate, high, and extreme. The results, presented in Table 3, demonstrate a marked degradation in segmentation accuracy as the noise level in the labels increases. Specifically, under high and extreme noise conditions, both the Dice Similarity Coefficient (DSC) and Normalized Surface Distance (NSD) metrics for all models exhibit substantial declines. For instance, the Dice Score for U-Net, nnU-Net, and SwinUNETR drops from 43.48, 47.41, and 43.09 at the low noise level to 32.49, 37.71, and 32.14 at the extreme noise level, respectively. These results underscore the critical importance of accurate synthetic tumor boundary generation for enhancing model performance. The findings highlight that maintaining high-fidelity synthetic data is essential to mitigate the adverse effects of label noise and improve the robustness and reliability of segmentation models in clinical applications.

Table 2. Performance comparison between original labels and noisy labels. We report the Dice Similarity Coefficient (DSC) and Normalized Surface Distance (NSD). Higher DSC and lower NSD indicate a superiority.

Method	DSC(%)↑	NSD(%)↓
With original labels		
U-Net	48.48	43.26
nnU-Net	50.50	47.32
SwinUNETR	45.88	39.76
With noisy labels		
U-Net	43.48	37.90
nnU-Net	47.41	41.78
SwinUNETR	43.09	38.30

Table 3. Overall performances of models with different levels of noisy labels. We can observe a marked degradation in segmentation accuracy as the noise level in the labels increases.

Method	Noise Level	DSC(%) ↑	NSD(%) ↓
With original labels			
U-Net	-	48.48	43.26
nnU-Net	-	50.50	47.32
SwinUNETR	-	45.88	39.76
With noisy labels			
U-Net	Low	43.48	37.90
	Moderate	40.86	35.31
	High	41.12	34.94
	Extreme	32.49	25.10
nnU-Net	Low	47.41	41.78
	Moderate	50.64	46.05
	High	44.38	38.81
	Extreme	37.71	32.55
SwinUNETR	Low	43.09	38.30
	Moderate	43.69	38.54
	High	40.94	34.80
	Extreme	32.14	26.42

4 Discussion and Conclusion

This work investigates the performance of leading deep learning segmentation models for pancreatic tumors. We leverage the *Difftumor* framework [4] to generate synthetic tumors and evaluate three established models: U-Net [19], nnU-Net [10], and SwinUNETR [7]. We hypothesize that incorporating synthetic tumors and refining their properties can improve segmentation accuracy. Our comprehensive experiments demonstrate that augmenting training data with synthetic tumors significantly enhances the models' ability to delineate tumor boundaries. Notably, the size of these synthetic tumors plays a critical role in segmentation performance. Furthermore, we investigate the impact of label noise on synthetic tumor boundaries. We demonstrate that accurate tumor boundary generation is another key to better segmentation performances

This study emphasizes the critical role of high-fidelity and well-controlled synthetic data for achieving superior segmentation results in pancreatic tumors. Our findings suggest that future research should focus on developing more sophisticated methods for generating synthetic data that closely resembles real-world pathological presentations. By achieving this level of realism, we can train models with enhanced segmentation capabilities, ultimately leading to improved applicability and effectiveness in clinical practice.

References

1. Antonelli, M., et al.: The medical segmentation decathlon. Nat. Commun. **13**(1), 4128 (2022)
2. Billot, B., et al.: SynthSeg: Segmentation of brain MRI scans of any contrast and resolution without retraining. Med. Image Anal. **86**, 102789 (2023)
3. Chen, M., Shi, X., Zhang, Y., Wu, D., Guizani, M.: Deep feature learning for medical image analysis with convolutional autoencoder neural network. IEEE Trans. Big Data **7**(4), 750–758 (2017)
4. Chen, Q., et al.: Towards generalizable tumor synthesis (2024). arXiv preprint arXiv:2402.19470
5. Esser, P., Rombach, R., Ommer, B.: Taming transformers for high-resolution image synthesis. In: Proceedings of the IEEE/CVF Conference on Computer Vision and Pattern Recognition, pp. 12873–12883 (2021)
6. Goodfellow, I., et al.: Generative adversarial networks. Commun. ACM **63**(11), 139–144 (2020)
7. Hatamizadeh, A., Nath, V., Tang, Y., Yang, D., Roth, H.R., Xu, D.: Swin unetr: swin transformers for semantic segmentation of brain tumors in mri images. In: International MICCAI Brainlesion Workshop, pp. 272–284. Springer (2021)
8. Ho, J., Jain, A., Abbeel, P.: Denoising diffusion probabilistic models. Adv. Neural. Inf. Process. Syst. **33**, 6840–6851 (2020)
9. Hu, Q., et al.: Label-free liver tumor segmentation. In: Proceedings of the IEEE/CVF Conference on Computer Vision and Pattern Recognition, pp. 7422–7432 (2023)
10. Isensee, F., Jaeger, P.F., Kohl, S.A., Petersen, J., Maier-Hein, K.H.: nnU-Net: a self-configuring method for deep learning-based biomedical image segmentation. Nat. Methods **18**(2), 203–211 (2021)
11. Islami, F., et al.: American cancer society's report on the status of cancer disparities in the united states, 2023. CA: A Cancer J. Clin. **74**(2), 136–166 (2024)
12. Jin, Q., Cui, H., Sun, C., Meng, Z., Su, R.: Free-form tumor synthesis in computed tomography images via richer generative adversarial network. Knowl. Based Syst. **218**, 106753 (2021)
13. Lai, Y., Chen, X., Wang, A., Yuille, A., Zhou, Z.: From pixel to cancer: Cellular automata in computed tomography (2024). arXiv preprint arXiv:2403.06459
14. Landman, B., Xu, Z., Igelsias, J., Styner, M., Langerak, T., Klein, A.: Miccai multi-atlas labeling beyond the cranial vault–workshop and challenge. In: Proc. MICCAI Multi-Atlas Labeling Beyond Cranial Vault-Workshop Challenge, vol. 5, p. 12 (2015)
15. Li, B., Chou, Y.C., Sun, S., Qiao, H., Yuille, A., Zhou, Z.: Early detection and localization of pancreatic cancer by label-free tumor synthesis (2023). arXiv preprint arXiv:2308.03008
16. Lyu, F., Ye, M., Carlsen, J.F., Erleben, K., Darkner, S., Yuen, P.C.: Pseudo-label guided image synthesis for semi-supervised covid-19 pneumonia infection segmentation. IEEE Trans. Med. Imaging **42**(3), 797–809 (2022)
17. Lyu, F., Ye, M., Ma, A.J., Yip, T.C.F., Wong, G.L.H., Yuen, P.C.: Learning from synthetic ct images via test-time training for liver tumor segmentation. IEEE Trans. Med. Imaging **41**(9), 2510–2520 (2022)
18. Rombach, R., Blattmann, A., Lorenz, D., Esser, P., Ommer, B.: High-resolution image synthesis with latent diffusion models. In: Proceedings of the IEEE/CVF Conference on Computer Vision and Pattern Recognition, pp. 10684–10695 (2022)

19. Ronneberger, O., Fischer, P., Brox, T.: U-net: convolutional networks for biomedical image segmentation. In: Medical image computing and computer-assisted intervention–MICCAI 2015: 18th international conference, Munich, Germany, October 5-9, 2015, proceedings, part III 18. pp. 234–241. Springer (2015)

20. Roth, H.R., Farag, A., Turkbey, E., Lu, L., Liu, J., Summers, R.M.: Data from pancreas-ct. the cancer imaging archive. IEEE Trans. Image Process. **10**, K9 (2016)

21. Shang, F., Fu, J., Yang, Y., Huang, H., Liu, J., Ma, L.: Synfundus: A synthetic fundus images dataset with millions of samples and multi-disease annotations (2023). arXiv preprint arXiv:2312.00377

22. Shen, Z., Ouyang, X., Xiao, B., Cheng, J.Z., Shen, D., Wang, Q.: Image synthesis with disentangled attributes for chest x-ray nodule augmentation and detection. Med. Image Anal. **84**, 102708 (2023)

23. Siegel, R.L., Miller, K.D., Wagle, N.S., Jemal, A.: Cancer statistics, 2023. CA: a cancer journal for clinicians **73**(1) (2023)

24. Wu, L., Zhuang, J., Ni, X., Chen, H.: Freetumor: Advance tumor segmentation via large-scale tumor synthesis (2024). arXiv preprint arXiv:2406.01264

25. Zhang, Z., Bagci, U.: Dynamic linear transformer for 3d biomedical image segmentation. In: International Workshop on Machine Learning in Medical Imaging, pp. 171–180. Springer (2022)

26. Zhang, Z., et al.: Large-scale multi-center ct and mri segmentation of pancreas with deep learning (2024). arXiv preprint arXiv:2405.12367

27. Zhang, Z., Yao, L., Keles, E., Velichko, Y., Bagci, U.: Deep learning algorithms for pancreas segmentation from radiology scans: a review. Adv. Clin. Radiol. **5**(1), 31–52 (2023)

Pancreatic Vessel Landmark Detection in CT Angiography Using Prior Anatomical Knowledge

Leonhard Rist[1,2(✉)], Christopher Homm[1], Felix Lades[2],
Abraham Ayala Hernandez[2], Michael Sühling[2], Erik Gudman Steuble Brandt[3],
Andreas Maier[1], and Oliver Taubmann[2]

[1] Friedrich-Alexander Universität, Erlangen-Nürnberg, Erlangen, Germany
leonhard.rist@fau.de
[2] CT R&D Image Analytics, Siemens Healthineers, Forchheim, Germany
[3] Department of Radiology, Herlev Gentofte Hospital, Copenhagen University
Hospital, Copenhagen, Denmark

Abstract. Localizing vessels or their defining bifurcations is a frequent clinical problem for advanced visualizations in pancreatic cancer invasion analysis, driving the demand for design guidelines of easy-to-implement landmark detection solutions. When transforming such landmarks appropriately to surrogate targets, the competitive nnDetection and nnU-Net frameworks provide such solutions, especially for small data settings. Here, the underlying networks can further benefit from incorporating additional anatomical information. We present results on two CTA datasets consisting of arterial and venous phase images with 6 and 4 bifurcation landmarks surrounding the pancreas respectively. Landmark points were modeled as spheres to allow for the application of object detection/segmentation models. We evaluate both nn-frameworks for these tasks focusing on the incorporation of anatomical knowledge. Postprocessing nnDetection predictions with organ masks and landmark relation constraints boosts detection accuracies from 66.7 % to 79.4 % in the more challenging venous case and decreases the mean radial error from 9.06 to 4.92 mm. The nnU-Net benefits more from organ masks in the input when targeting problematic vessels, lowering the mean radial error from 12.97 to 8.45 mm when using the splenic mask for the venous task. Both networks have good initial detection rates for the arterial phase which are slightly boosted using our method to 93.7 % (nnU-Net) and 95.5 % (nnDetection). All remaining mispredictions are within the vessel of interest and thus sufficient for many downstream tasks.

Keywords: mCTA · Landmark Detection · Vascular · Pancreas

1 Introduction

Diagnosing complex diseases such as pancreatic cancer can benefit from Deep Learning in multiple ways, typically ranging from organ segmentation [20] to

---

Leonhard Rist , Christopher Homm : both authors contributed equally to this work.

© The Author(s), under exclusive license to Springer Nature Switzerland AG 2025
F. Proietto Salanitri et al. (Eds.): PILM 2024/AIPAD 2024, LNCS 15197, pp. 45–54, 2025.
https://doi.org/10.1007/978-3-031-73483-0_5

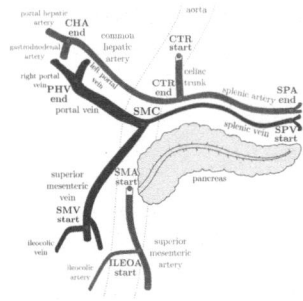  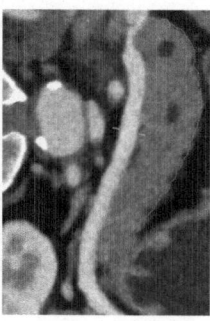

(a) Schematic view of all bifurca-   (b) Pancreas and surrounding   (c) Unfolded splenic
tions of interest   vessels in 3D   artery with pancreas

Fig. 1. Pancreas and surrounding vessels of interest in a schematic and 3D view (left, center), next to an unfolded CTA image.

outcome prediction [24]. However, for treatment planning, the assessment must also include a comprehensive investigation of cancer infestation in the surrounding vasculature that is shown in Fig. 1a. Hence, reliably extracting vessel segments is crucial [4] to allow advanced centerline-aligned reformations and visualizations, see Figs. 1c & 1b. This task can be simplified to finding bifurcations first, followed by applying vessel tracing algorithms. However, even then, the effective training of these algorithms is often limited by the necessity of hyperparameter tuning or improper use of existing anatomical knowledge.

Classical landmark detection approaches such as Active Shape Models [6] or Histograms of Oriented Gradients [7] do not translate well to 3D Computed Tomography Angiography (CTA) data. Convolutional neural networks which directly regress coordinates [12] [21] mostly use 2D data, failing to capture 3D context. The complex final coordinate mapping also renders these approaches less competitive. Promising landmark detection approaches employ reinforcement learning, such as training agents for image navigation by Ghesu et al. [9], or multi-stage methods such as Kang et al. [15]. State-of-the-art heatmap regression models, also often for 2D, predict heatmap distance maps, potentially using transformers and multi-resolution [23] or attention [25]. For trainable end-to-end extraction, Nibali et al. [18] introduced Differentiable Spatial to Numerical Transforms which have been adapted to the medical domain [8]. Object detectors are designed as two-stage networks such as Mask [11] or Faster R-CNN (e.g. [5]) or one-stage such as Retina U-Net [14]. Finally, graph networks can be used to refine heatmap or detector predictions [16] [17].

However, many of these approaches are non-trivial in setup and training. Instead, for less involved architectures or common segmentation models [2], the problem can be simplified to defining surrogate objects around the landmarks. Additionally, the rise of self-configuring frameworks enables high-performing results even with little data and simplifies the reproduction of results. Most prominent are the nnU-Net [13] for segmentation and the nnDetection [3] for object detection tasks, which are feasible for this task when reformulating the problem using surrogate targets. Even the latest specialized architectures, such as various Vision Transformers [10], often cannot improve their performance in real-world medical scenarios due to their high data requirements. Additionally, available prior knowledge (neighboring organs, relationship between landmarks etc.) is often neglected while possibly providing chances of improving results compared to hyperparameter tweaking.

Focusing on applications around the pancreas, this paper thrives to give guidelines for how to incorporate existing anatomical knowledge in simple ways, contributing as follows: In the example of 6 arterial and 4 venous vascular landmarks surrounding the pancreas in 2 different phase CTA datasets, we show how such bifurcation detection tasks can be transformed into problems best solved by nn-frameworks, including design questions such as the size of the surrogate object. We further suggest how to enforce constraints for landmark relations in addition to different means of incorporating organ masks in the input or in post-processing to remove outliers. We present our results using the nn-frameworks, to be independent of specialized architectural design choices and focus on the effect of the presented anatomical priors which can be adapted by basically any architecture. In our experiments, these simple adaptations already lead to predictions that are always inside the vascular structure of interest.

2 Methods

2.1 Datasets

The task at hand is to retrieve 6 arterial and 4 venous landmarks from abdominal or whole-body CTA scans. All landmarks are start and end points for vessels surrounding the pancreas as depicted and defined in Fig. 1a. The used datasets consist of 246 scans acquired in the arterial phase and 156 scans in the venous phase. Of these, 274 scans were acquired using low-noise photon-counting CT technology, improving the visibility of small structures in the pancreas. Only data with a slice thickness of up to 2 mm were used and resampled to a spatial resolution of 0.7 mm×0.7 mm×0.6 mm. The arterial and venous bifurcations were labeled in the scans of their respective contrast phase only. Two medical annotators provided annotations which were then unified by a third annotator following a defined rule set regarding the vascular topology. The scans were collected from multiple retrospective studies and each received Institutional Review Board approval prior to starting. The need for consent was either waived or given. The train/validation/test split was set to 172/43/31 scans for the arterial and to 115/26/15 for the venous landmarks.

More and more, automatically computed organ masks [1] and standard landmarks [9] are available from existing clinical tools. Hence, to narrow down the search space and eliminate the need for multi-stage detectors, we make use of common organ and bone landmarks to roughly crop the scans to an abdominal region around the pancreas as shown in the red squares inside the scans in Fig. 2. A reasonable set of such landmarks can quickly be found by filtering for

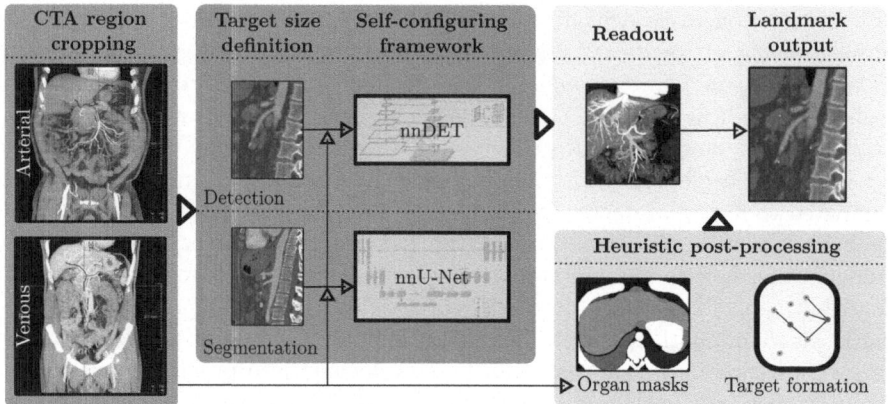

Fig. 2. Stages where anatomical context is included in the pipeline.

landmarks which best surround the bifurcations in the training set and applying small additional margins to build the cropped volume. This eliminates possibly misleading vessel structures and decreases training and inference times.

2.2   Models and Problem Design

Automatic configuration of the established 3-D U-Net segmentation model [19] in nnU-Net [13] and of Retina-Net detection model [14] in nnDetection [3] increased reproducibility and applicability for non-experts and often outperforms other architectures. We could confirm this when comparing initial results to a SWIN UNETR Transformer [22] (pre-trained on CT images). Note, that the included anatomical priors (additional input or post-processing) are not architecture bound, allowing a transfer to state-of-the-art heatmap regression models. Since we target easy-to-setup pipelines, only the nn-frameworks are compared in this work.

The nnU-Net requires semantic segmentation masks while nnDetection requires instance masks. We suggest using spherical objects as surrogates centered on the landmark positions. Their radius is initially chosen to be 5 mm inspired by the vessel radii of interest. Both models use the centroid of their predictions as final landmark positions. Specifically, for the segmentation masks the centroid of the largest connected component is used (after 80 % maximum thresholding).

2.3   Incorporating Anatomical Knowledge

Anatomical knowledge which is easy to access and incorporate consists mainly of organ masks or additional standard landmarks as they can be directly acquired from the scan and are tools likely to be at hand. Furthermore, since landmarks are predicted simultaneously and can consist of multiple candidates, one can

compare their geometric relations and only choose anatomically plausible formations. Hence, we use these geometrical formations for post-processing. Individual
organ masks, however, can either be incorporated directly into the input (e.g. as
additional channels) or used as post-processing filters, see Fig. 2. The abdominal
volumetric organ masks used in this work are retrieved from a commercial tool
[1]. As nnDetection produces multiple candidates, while nnU-Net optimizes for
single components, the former is expected make better use of post-processing
while the latter might benefit more from extra input.

For vascular bifurcation detection around the pancreas, masks of the closest organs—spleen, liver, pancreas, and both kidneys—are investigated as input.
They are either fed individually as one-hot encoded channels or in the case of
multiple organs in a single channel (equidistant, contiguous values). Chosen landmark formation constraints should be robust against likely changes in patient
position. The pancreatic bifurcations of interest show consistent topological relationships, hence this work opts for their direct exploitation instead of employing
more elaborate methods such as Markov Random Fields. Topological constraints
are executed subsequently ranked by their importance for the task and are either
global (e.g., most superior landmark) or bi/tri-relational (A always superior to
B), leading to the discarding of implausible landmark candidates. For this specific task, we chose at least one constraint per landmark. Prominent constraints
are: On a global level, *ILEOA start* must be most inferior, *SPV start* most left,
SMV start most superior, *SMC* longitudinal between *PHV end* and *SMV start*.

2.4 Experiments

We report the effect of landmark formation constraints, of organ masks in input
and post-processing, and of spherical target sizes. All experiments are evaluated
on the arterial and venous datasets using both nnU-Net and nnDetection. As
some design choices are more relevant for only one of the architectures, only
relevant choices are reported for brevity. Metrics are reported for the test set.

Prediction quality is measured using the radial error ERR for each landmark,
i.e. the Euclidean distance from the extracted to the annotated reference landmark. We further state a detection accuracy ACC_t which counts the bifurcations
as detected only within a distance of $t = 5\,mm$. For bigger junction areas, namely
aorta (25–35 mm in diameter) and *SMC* junction, t is chosen to be 10 mm.

3   Results

Baseline The baseline results, shown in Table 1, indicate that nnU-Net with a
mean (median) ERR of 3.71 mm and a detection rate of 93.6 % and nnDetection
with 3.3 mm and 94.2 % have a decent initial performance on the arterial data.
This highlights a sufficient capability of the nn-frameworks for this task.

To note, nnU-Net has close to 100 % normalized detection accuracy for each
bifurcation except for *SPA end* with 61.5 % even though its median radial error
of 3.1 mm is comparable to all other vessels (2.6 - 3.7 mm). Accuracy scores for

Table 1. Overall results, comparing the baselines to the best combinations, meaning post-processing for nnDetection and spleen mask plus post-processing for nnU-Net. Radial errors are in mm, detection accuracies in %.

	Arterial				Venous			
	nnDet		nnU-Net		nnDet		nnU-Net	
	ACC_t	ERR	ACC_t	ERR	ACC_t	ERR	ACC_t	ERR
baseline	94.2	3.30	93.6	3.71	66.7	9.06	60.0	11.33
best	**95.5**	**2.95**	93.7	3.13	**79.4**	**4.92**	68.7	8.06

Table 2. Most influential methods per framework: post-processing with organ masks (OM) or landmark formation (LMF), or using input masks (mask). For nnU-Net the input mask in the arterial phase contains 5 organs in one channel and the spleen mask only for the venous phase. Baselines show absolute, ablations relative values in mm.

	nnDetection				nnU-Net		
	ERR base	OM	LMF	OM+LMF	ERR base	OM+LMF	mask
arterial	3.30	± 0	−0.35	−0.35	3.71	−0.96	−0.65
venous	9.06	−0.69	−2.74	**−4.14**	11.33	−0.86	**−2.88**

nnDetection are more balanced with *SPA end* being detected in 80.8 % of the cases. For the venous phase, performances for both models drop. Plain nnU-Net results in a 60.0 % detection accuracy with a mean (median) ERR of 11.33 mm (4.63 mm) and nnDetection achieves 66.7 % detection accuracy with 9.06 mm (3.48 mm) error. This is mostly due to *SPV start* being detected in only 20 % and 26 % of the cases for nnU-Net and nnDetection, with a higher mean (21.9 mm and 19.9 mm) and standard deviation in the radial error, see Fig. 4. Considering this, the subsequent focus lies on outlier removal and *SPV start*.

Heuristic post-processing The post-processing methods: organ mask filtering, enforcing landmark constraints and their combination respectively improves the mean radial error from 3.71 to 2.75 mm in the arterial phase for the nnU-Net, see combination in Table 2. The combination of both methods improves venous predictions from 11.33 to 10.47 mm. For nnDetection, the mean error decreases from 3.30 to 2.95 mm for the arterial bifurcations, only benefiting from the landmark formation, and from 9.06 to 4.92 mm using both methods in the venous phase. For the latter, using landmark formation or organ masks individually improves the mean ERR to 6.32 and 8.37 mm respectively. The post-processing improves all venous landmarks, most importantly shrinking the median radial error of *SPV start* from 19.9 to 5.3 mm, see Fig. 3a.

Input Organ Masks. Since the heuristic post-processing already solves mispredictions located in neighboring organs in all cases for the nnDetection, no further evaluation was indicated in this case. For nnU-Net, additional input masks have a different effect for both contrast phases. Most relevantly, no one-hot encoded

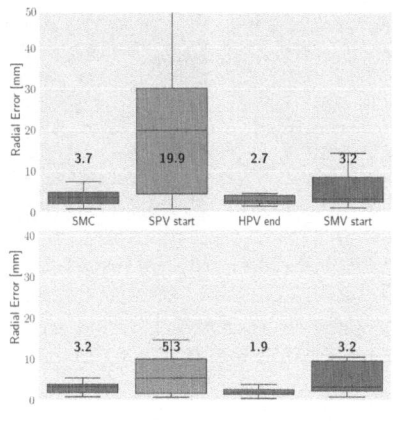

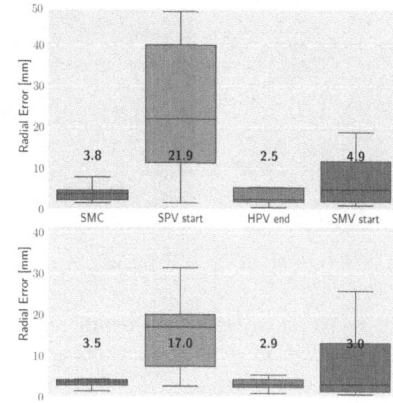

(a) nnDetection with OM+LMF   (b) nnU-Net with spleen mask

Fig. 3. Radial errors of the venous landmarks: Baseline is on top, improvement below. Boxes show median and range from 2nd to 3rd quartile.

channel mask (combination) could improve the segmentation baseline. However, combining the 5 anatomically close masks in a single channel slightly improved the mean radial error by 0.92 mm, however with no clear trend w.r.t the individual landmarks. This approach did not improve the mean results for the venous case (14.99 mm). However, specifically targeting the difficult splenic bifurcation *SPV start*, mask combinations using the spleen improved the results. The spleen mask alone performed best, reducing the mean error to 8.45 mm (21.9 to 17.0 mm median error for *SPV start*, along with reduced standard deviation), see Fig. 3b.

Target Sizes. For the segmentation task, too small radii (e.g. 3 mm) do not lead to convergence. Increasing the size, however, from 5 mm to 7 mm and 9 mm, consistently increases the detection performance from 93.6 over 95.0 to 95.5 % (reducing the mean radial error by 1.3 mm) in the arterial case and from 60.0 over 61.7 to 62.6 % (reducing mean error by 2.83 mm) in the venous case. The detection model does converge using a 3 mm radius; increasing the target size gives varying results compared to the baseline with no clear trend.

4   Discussion and Conclusion

This work focuses on incorporating anatomical information for landmark regression in pancreatic vascular analysis which is tackled with self-configuring network frameworks. The method itself can be applied independently of the architectural choice. In this setup, we also revisit the target transformation into segmentation or object detection tasks. Evaluations are based on two abdominal CTA datasets from two different contrast phases for the application of pancreatic vessel bifurcation detection, giving valuable information when assessing pancreatic diseases and treatment plans. Cropping these scans to a rough ROI utilizing commonly

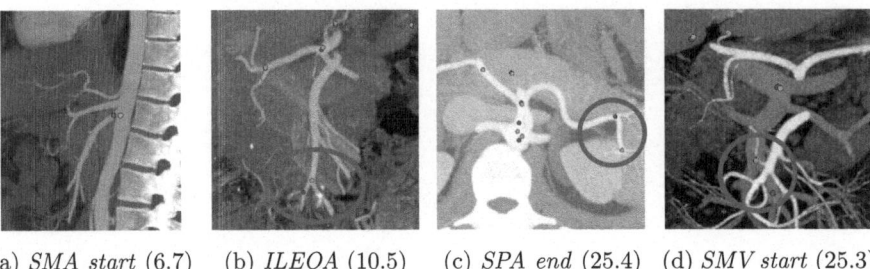

(a) *SMA start* (6.7)   (b) *ILEOA* (10.5)   (c) *SPA end* (25.4)   (d) *SMV start* (25.3)

Fig. 4. Worst nnDetection results with post-processing for each vessel (radial error in mm) on MIP images. Predictions are green, Ground Truth is red. (Color figure online)

available landmarks rendered more complex multi-stage approaches unnecessary while initial results discouraged the use of data-hungry Transformer architectures. The achieved mean and median radial errors are (except for 8.06 mm in nnU-Net for the venous data) smaller than the real radii at the targeted vessel bifurcations (approx. 5 mm radius) and all remaining outliers lie within the correct vessel as shown in Fig. 4, rendering this solution exceptionally robust for vessel tracing tasks. Furthermore, when investigating predictions with small radial errors, meaning located at the correct bifurcation, our predictions are often placed more intuitively compared to the manual annotation (Fig. 4a).

Both nnU-Net and nnDetection are able to solve this task and benefit from anatomical knowledge integration in different ways, with nnDetection exhibiting slightly better performance. Prior knowledge about the mutual landmark relations or using surrounding structures can be beneficially incorporated in two ways depending on the architecture choice. For the segmentation model, which mostly outputs a single prediction component, it is beneficial to incorporate such information as input masks when problematic landmarks can be targeted directly (spleen mask for the splenic vein). Both models had problems when handling many organ masks at once in the input. For detection tasks where multiple bounding boxes with confidence scores are available, these organ masks can be used more directly by filtering out predictions. Additionally, since the landmarks of a patient are correlated, enforcing certain geometric relations on the predictions rules out all remaining anatomically implausible predictions for the detection model. In other scenarios, where relationships might not be as clearly defined as in this use case, an additional graph layer or Markov Random Field might help to find the best overall configuration. While the detection model does not have a clear preference for its spherical target size, the segmentation task benefits from larger targets (more similar to heatmap extents) and does not converge for too small sizes (3 mm), possibly due to insufficient loss feedback.

Next to demonstrating very good performance detecting pancreas-relevant vasculature in CTA, these insights are also intended to help readers design their own pipelines for other landmark detection tasks without the need for highly specialized architectures and hyperparameter tuning. Providing simple guide-lines for incorporating prior knowledge increases reliability and diminishes the black-box character of Deep Learning approaches, encouraging their application in practice. For the task at hand, this resulted in the complete elimination of out-liers outside of the vessels of interest compared to the unmodified nn-baselines.

References

1. A deep image-to-image network organ segmentation algorithm for radiation treat-ment planning: principles and evaluation. Radiat. oncol. (London, England) **17**(1), 129 (2022). https://doi.org/10.1186/S13014-022-02102-6
2. Azad, R., et al.: Medical Image Segmentation Review: the success of U-Net (11 2022). https://arxiv.org/abs/2211.14830v1
3. Baumgartner, M., Jäger, P.F., Isensee, F., Maier-Hein, K.H.: nnDetection: a self-configuring method for medical object detection. In: de Bruijne, M., et al. (eds.) Medical Image Computing and Computer Assisted Intervention - MICCAI 2021, pp. 530–539. Springer International Publishing, Cham (2021). https://doi.org/10.1007/978-3-030-87240-3_51
4. Buchs, N.C., Chilcott, M., Poletti, P.A., Buhler, L.H., Morel, P.: Vascular invasion in pancreatic cancer: imaging modalities, preoperative diagnosis and surgical man-agement. World J. Gastroenterol. WJG **16**, 818 (2010). https://doi.org/10.3748/WJG.V16.I7.818
5. Chen, X., et al.: Fast and accurate craniomaxillofacial landmark detection via 3D faster R-CNN. IEEE Trans. Med. Imaging **40**, 3867–3878 (2021). https://doi.org/10.1109/TMI.2021.3099509
6. Cootes, T., Taylor, C., Cooper, D., Graham, J.: Active shape models-their training and application. Comput. Vis. Image Underst. **61**(1), 38–59 (1995). https://doi.org/10.1006/cviu.1995.1004
7. Dalal, N., Triggs, B.: Histograms of oriented gradients for human detection. In: 2005 IEEE Computer Society Conference on Computer Vision and Pattern Recognition (CVPR'05), vol. 1, pp. 886–893 vol. 1 (2005).https://doi.org/10.1109/CVPR.2005.177
8. Gajowczyk, M., et al.: Coronary ostia localization using residual U-Net with heatmap matching and 3D DSNT. In: International Workshop on Machine Learn-ing in Medical Imaging, LNCS, pp. 318–327 (2022)
9. Ghesu, F.C., Georgescu, B., Zheng, Y., Grbic, S., Maier, A., Hornegger, J., Comani-ciu, D.: Multi-scale deep reinforcement learning for real-time 3D-landmark detec-tion in CT scans. IEEE Trans. Pattern Anal. Mach. Intell. **41**, 176–189 (2019). https://doi.org/10.1109/TPAMI.2017.2782687
10. Han, K., et al.: A survey on vision transformer. IEEE Trans. Pattern Anal. Mach. Intell. **45**, 87–110 (2023). https://doi.org/10.1109/TPAMI.2022.3152247
11. He, K., Gkioxari, G., Dollar, P., Girshick, R.: Mask R-CNN. In: Proceedings of the IEEE International Conference on Computer Vision (ICCV) (Oct 2017)
12. Huang, Y., Liu, X., Jin, L., Zhang, X.: DeepFinger: a cascade convolutional neuron network approach to finger key point detection in egocentric vision with mobile camera. In: Proceedings - 2015 IEEE International Conference on Systems, Man, and Cybernetics, SMC 2015, pp. 2944–2949 (2016). https://doi.org/10.1109/SMC.2015.512

13. Isensee, F., Jaeger, P.F., Kohl, S.A.A., Petersen, J., Maier-Hein, K.H.: nnU-Net: a self-configuring method for deep learning-based biomedical image segmentation. Nat. Methods **18**(2), 203–211 (2021). https://doi.org/10.1038/s41592-020-01008-z

14. Jaeger, P.F., et al.: Retina U-Net: embarrassingly simple exploitation of segmentation supervision for medical object detection. In: Proceedings of the Machine Learning for Health NeurIPS Workshop. Proceedings of Machine Learning Research, vol. 116, pp. 171–183. PMLR (13 Dec 2020)

15. Kang, S.H., Jeon, K., Kang, S.H., Lee, S.H.: 3D cephalometric landmark detection by multiple stage deep reinforcement learning. Sci. Rep. **11**, 17509 (2021). https://doi.org/10.1038/s41598-021-97116-7

16. Lang, Y., et al.: Automatic localization of landmarks in craniomaxillofacial CBCT images using a local attention-based graph convolution network. Med. Image Comput. Comput. Assist. Interv. - MICCAI **2020**(12264), 817–826 (2020)

17. Nguyen, L.Q., et al.: Facial landmark detection with learnable connectivity graph convolutional network. IEEE Access (2022). https://doi.org/10.1109/ACCESS.2022.3200037

18. Nibali, A., He, Z., Morgan, S., Prendergast, L.: Numerical coordinate regression with convolutional neural networks (2018). https://arxiv.org/abs/1801.07372v2

19. Ronneberger, O., Fischer, P., Brox, T.: U-net: convolutional networks for biomedical image segmentation. In: Medical Image Computing And Computer-assisted Intervention–MICCAI 2015: 18th International Conference, Munich, Germany, October 5-9, 2015, proceedings, part III 18 , LNCS, vol. 9351, pp. 234–241 (2015).https://doi.org/10.1007/978-3-319-24574-4_28/COVER

20. Roth, H.R., Lu, L., Farag, A., Shin, H.C., Liu, J., Turkbey, E.B., Summers, R.M.: DeepOrgan: multi-level deep convolutional networks for automated pancreas segmentation. In: Navab, N., Hornegger, J., Wells, W.M., Frangi, A. (eds.) Medical Image Computing and Computer-Assisted Intervention - MICCAI 2015, pp. 556–564. Springer International Publishing, Cham (2015). https://doi.org/10.1007/978-3-319-24553-9_68

21. Tack, A., Preim, B., Zachow, S.: Fully automated assessment of knee alignment from full-leg X-Rays employing a "YOLOv4 And Resnet Landmark regression Algorithm" (YARLA): data from the osteoarthritis Initiative. Comput. Methods Programs Biomed. **205**, 106080 (2021). https://doi.org/10.1016/J.CMPB.2021.106080

22. Tang, Y., Yang, D., Li, W., Roth, H.R., Landman, B., Xu, D., Nath, V., Hatamizadeh, A.: Self-supervised pre-training of swin transformers for 3D medical image analysis. In: Proceedings of the IEEE/CVF Conference on Computer Vision and Pattern Recognition, pp. 20730–20740 (2022)

23. Viriyasaranon, T., Ma, S., Choi, J.H.: Anatomical landmark detection using a multiresolution learning approach with a hybrid transformer-CNN model. In: Greenspan, H., et al. (eds.) Medical Image Computing and Computer Assisted Intervention - MICCAI 2023, pp. 433–443. Springer Nature Switzerland, Cham (2023). https://doi.org/10.1007/978-3-031-43987-2_42

24. Yokoyama, S., Hamada, T., Higashi, M.: Predicted prognosis of patients with pancreatic cancer by machine learning. Clin. Cancer Res. **26**(10), 2411–2421 (2020).https://doi.org/10.1158/1078-0432.CCR-19-1247

25. Zhong, Z., Li, J., Zhang, Z., Jiao, Z., Gao, X.: An attention-guided deep regression model for landmark detection in cephalograms. In: Shen, D., et al. (eds.) Medical Image Computing and Computer Assisted Intervention - MICCAI 2019, pp. 540–548. Springer International Publishing, Cham (2019)

Personalized Incremental Learning
in Medicine

Addressing Catastrophic Forgetting by Modulating Global Batch Normalization Statistics for Medical Domain Expansion

Sharut Gupta[1,2], Ken Chang[3], Liangqiong Qu[4], Aakanksha Rana[2],
Syed Rakin Ahmed[5], Mehak Aggarwal[1], Nishanth Arun[6], Ashwin Vaswani[6],
Shruti Raghavan[7], Vibha Agarwal[2], Mishka Gidwani[1], Katharina Hoebel[5],
Jay Patel[1], Charles Lu[2], Christopher P. Bridge[1], Daniel L. Rubin[3],
Jayashree Kalpathy-Cramer[8], and Praveer Singh[8(✉)]

[1] Massachusetts General Hospital, Boston, MA, USA
[2] Massachusetts Institute of Technology, Cambridge, MA, USA
[3] Stanford University, Palo Alto, CA, USA
[4] The University of Hong Kong, Pok Fu Lam, Hong Kong
[5] Harvard Medical School, Boston, MA, USA
[6] Carnegie-Mellon University, Pittsburgh, PA, USA
[7] The University of Texas at Austin, Austin, TX, USA
[8] University of Colorado School of Medicine, Aurora, CO, USA
praveer.singh@cuanschutz.edu

Abstract. Model brittleness across datasets is a key concern when deploying deep learning models in real-world medical settings. One approach is to fine-tune the model on subsequent datasets after training on the original dataset. However, this degrades model performance on the original dataset, a phenomenon known as *catastrophic forgetting*. We develop an approach to address catastrophic forgetting by combining elastic weight consolidation with a simple yet novel modulation of global batch normalization statistics under two scenarios: expanding the domain across 1) imaging systems and 2) hospital institutions. Focusing on the clinical use case of mammographic breast density detection, we show that our approach empirically outperforms several other state-of-the-art approaches and provides theoretical justification for the efficacy of batch normalization modulation, demonstrating the potential of our approach to deploying clinical deep learning models requiring domain expansion.

Keywords: Catastrophic Forgetting · Deep Learning · Domain Expansion

---

S. Gupta and K. Chang—Co-first authors.

J. Kalpathy-Cramer and P. Singh—Co-senior authors.

© The Author(s), under exclusive license to Springer Nature Switzerland AG 2025
F. Proietto Salanitri et al. (Eds.): PILM 2024/AIPAD 2024, LNCS 15197, pp. 57–72, 2025.
https://doi.org/10.1007/978-3-031-73483-0_6

1   Introduction

Deep learning (DL) models have shown state-of-the-art performance for a wide variety of computer vision [2,24] and medical imaging tasks [3,11]. Within the clinical context, there is a need to continually refine these models to achieve high performance on new datasets, such as those from different image acquisition systems or hospitals. Typically, DL models are extended to new datasets via fine-tuning the model weights; a neural network trained on the original dataset is used to initialize a new model, which is then trained on the target domain [17]. However, this can result in *catastrophic forgetting* (CF) of the previous dataset, a phenomenon in which models do not preserve previously learned knowledge and consequently result in a degradation of the performance on the original dataset [4]. This poses a major challenge for regulatory agencies, such as the Food and Drug Administration (FDA) in the United States. DL models that have been fine-tuned after their approval may no longer satisfy the required performance criteria on the original test set. In this work, we explore techniques to mitigate CF in the setting of fine-tuning on new medical datasets, called domain expansion.

Prior work in addressing CF in domain expansion has focused on interventions that target specific layers of a network, *e.g.*, by either fine-tuning trainable parameters of the Batch Normalization (BN) layers [7] or modulating the BN statistics (bias/variance) [12]; both of them attempting to rectify any differences in the internal co-variate shift by aligning the distribution of the new dataset with the previous one, resulting in similar model performance across both datasets. However, these techniques limit the model's capacity to incorporate new knowledge since all non-BN layers are frozen throughout the fine-tuning process.

Recent work [21,25] has shown that the creation of site-specific models can potentially recover performance loss from CF. However, individualized models are more complicated to both train and deploy than a single, robust model. Other approaches specifically handle CF by constraining gradient updates to model parameters. For example, Elastic Weight Consolidation (EWC) [10] updates parameters proportionally to the inverse of each parameter's importance to the original training dataset (*i.e.*, the magnitude of updates for more important parameters is small). Zeng et al. [28] claim EWC to be ineffective in retaining performance on the original task and instead, propose Orthogonal Weight Modification (OWM) wherein parameters are updated in a direction orthogonal to the subspace spanned by the inputs of the model. However, a major limitation of OWM in medical applications is that OWM requires the underlying model to first perform feature extraction from the combined dataset before training a multi-layered perceptron on top of the extracted features. This is usually impractical in our problem setting where datasets between multiple hospitals cannot be combined at a central location due to infrastructural & patient privacy issues.

Unlike previous work, we focus on the development of techniques to address CF in a more *realistic setting*. We consider a real-world clinical application of mammographic breast density assessment, which is routinely used to assess breast cancer risk to decrease the chances of breast cancer mortality [19]. Specifically, the identification of patients with dense breast tissue warrants additional monitoring, such as supplemental ultrasound or magnetic resonance imaging. The current criteria for mammographic breast density classification is based on the Breast Imaging Reporting and Data System (BI-RADS), which divides breast density into four distinct categories: fatty, scattered, heterogeneously dense, and extremely dense [13]. Given the variability of manual interpretation, there has been interest in developing automated approaches for the assessment of mammographic density [23]. A major hurdle for large-scale clinical deployment of a deep learning-based breast density assessment tool is poor generalizability across different hospitals/institutions and scanner types owing to inherent variability in patient demographics, disease prevalence, and imaging acquisition techniques [1,27]. Our study addresses heterogeneity in digital mammography systems across different institutions (Fig. 1a)) that arises from variability in x-ray tube targets, filters, digital detector technology, and control of automatic exposure [8].

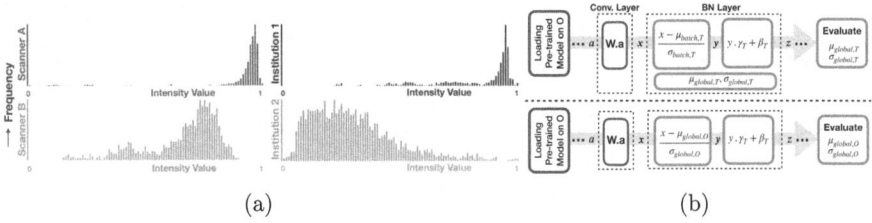

(a)   (b)

Fig. 1. (a)Intensity histograms showing heterogeneity across digital mammography systems (Scanner A vs. Scanner B) as well as across medical institutions (Institution 1 v/s Institution 2) (b) The two different BN approaches for fine-tuning on Dataset T: (Top) Using global BN statistics of T ($\mu_{global,T}$ & $\sigma_{global,T}$ computed during the fine-tuning process); (Bottom) Using global BN statistics of O ($\mu_{global,O}$ & $\sigma_{global,O}$ computed when originally training the model on O).

We investigate domain expansion techniques across different digital mammography systems (Scanner A & B) and institutions (Institutions 1 & 2) with an objective to mitigate CF.

For simplicity, the original domain (Scanner A, Institution 1), is referred to as *Dataset O* while the target domain (Scanner B, Institution 2) is referred to as *Dataset T*. The key contributions are as follows:

- We propose a simple yet effective technique to mitigate CF by utilizing global BN statistics (running mean & standard deviation of BN layers computed when training on that dataset) of Dataset O instead of Dataset T, when fine-tuning on T.
- We demonstrate the efficacy of this technique under two different scenarios: first when restricting fine-tuning to only BN layers (motivated by [7]) and second when fine-tuning using all the layers.

- We demonstrate how a commonly-used continuous learning algorithm (EWC [10]) fails for large, real-world datasets and further highlight how augmenting EWC with our technique not only improves domain expansion over T but also CF on O.
- By analyzing the probability density of per-channel activation and visualizing the UMAPs of features, we demonstrate the advantages of using the global batch normalization statistics of Dataset O.
- Lastly, we provide a theoretical justification for why using global BN statistics of Dataset O instead of T better mitigates CF.

2   Methods

2.1   Global BN Statistics of Dataset T vs. Dataset O

BN has become a popular technique to regulate shifts in the distribution of network activations during training [6]. BN normalizes each batch of input to the network layers by subtracting the batch mean and dividing by the batch standard deviation. Figure 1b) (top row) illustrates pipeline for traditional fine-tuning on T starting from a pre-trained model on O. During fine-tuning, output from the previous convolutional layer is passed to the BN layer, which normalizes the input $x_1, x_2...x_m$ (where m is the batch size):

$$y_i = \gamma_T \frac{x_i - \mu_{batch,T}}{\sqrt{\sigma_{batch,T}^2 + \epsilon}} + \beta_T, \tag{1}$$

where y_i is the output after applying BN on input x_i, $\mu_{batch,T}$ and $\sigma_{batch,T}^2$ are the batch mean and batch variance of T, respectively, ϵ is a stability parameter, and γ_T and β_T are trainable parameters of BN layers when fine-tuned with T. During inference, BN uses the global BN statistics of T (running mean $\mu_{global,T}$ and the running standard deviation $\sigma_{global,T}$) which were calculated while training; specifically, at training iteration k, the running mean $\mu_{global,T}[k]$ and running variance $\sigma_{global,T}^2[k]$ for T are updated:

$$\mu_{global,T}[k] = \alpha\mu_{global,T}[k-1] + (1-\alpha) * \mu_{batch,T}[k]$$
$$\sigma_{global,T}^2[k] = \alpha\sigma_{global,T}^2[k-1] + (1-\alpha) * \sigma_{batch,T}^2[k]$$

where the default momentum parameter, α, is set at 0.9. During inference, we replace $\mu_{batch,T}$ and $\sigma_{batch,T}^2$ in Eq. 1 with the global BN statistics of T computed during training, resulting in:

$$y_i = \gamma_T \frac{x_i - \mu_{global,T}}{\sqrt{\sigma_{global,T}^2 + \epsilon}} + \beta_T \tag{2}$$

In addition to the traditional approach to compute global BN statistics ($\mu_{global,T}$ and $\sigma_{global,T}$) fine-tuned on T and evaluated with the same global BN statistics for both datasets, we also conduct experiments to both fine-tune and evaluate using global BN statistics of O (as shown in bottom row of Fig 1b)). Thus, the BN (both for fine-tuning & inference) on an input batch can be represented by:

$$y_i = \gamma_T \frac{x_i - \mu_{global,O}}{\sqrt{\sigma_{global,O}^2 + \epsilon}} + \beta_T \qquad (3)$$

where $\mu_{global,O}$ and $\sigma_{global,O}$ are the global BN statistics of O (running mean and running standard deviation computed when training the model originally on O); γ_T and β_T denote the trainable parameters of BN layers when fine-tuning with T but using global BN statistics of O.

We experiment with fine-tuning the model using the above two global BN statistics techniques under the following two scenarios:

- **Fine tuning only BN layers**: As shown in Fig 1a), the intensity distributions of O (Scanner A in red and Institution 1 in blue) differ considerably from T (Scanner B in orange and Institution 2 in green), despite all being mammography screening datasets. With the intuition of aligning the data distribution of T with O, and retaining the original model performance for O, we first test run several experiments by fine-tuning only BN layers on T while freezing all other layers.
- **Fine-tuning all layers (BN and convolutional layers)**: With the primary objective of achieving the best performance on T while retaining the original performance on O, we further investigate the effects of fine-tuning all layers. Training using all layers aids in leveraging the full capacity of a model for fine-tuning on T.

2.2   Incorporation of EWC

While fine-tuning a model (previously trained on O) on T, EWC [10] effectively constrains the updates of those weights that are most important for O, thus allowing the model to converge to a minimum close to global minima of both T and O and ultimately prevents CF for O.

We implement this constraint using a quadratic penalty on the change in model parameters. To identify the salient parameters for model trained on O, we first compute the empirical Fisher Information Matrix (FIM). [14] FIM is defined as the covariance of the score function used to calculate the quality of parameter estimation. Due to the complexity of the likelihood function, computing this expectation becomes intractable and hence an approximation called the empirical Fisher (F) is defined as $F = \frac{1}{N} \sum_{i=1}^{N} \nabla_\theta \log p(x|\theta) \nabla_\theta \log p(x|\theta)'$.

This empirical Fisher matrix when combined with the quadratic penalty adds a constraint to the important parameters of model trained on O. The combined loss function $L(\theta)$ for elastic weight consolidation is given by $L(\theta) = L_T(\theta) + \sum_i \frac{\lambda}{2} F_i (\theta_i - \theta_{O,i}^*)^2$

where $L_T(\theta)$ is the loss function for T, θ_O^* represents the optimal model parameters for O and i in θ_i iterates over all the current model parameters which are being fine-tuned on T. The parameter λ is used as a trade off between the relative importance of performance over O and T (the higher the λ parameter, the closer will be the performance of the fine-tuned model to the model originally trained on Dataset O).

As a result, a constraint on some (not all) model parameters, allows the solution to stay in a low-error region which is optimal for both the datasets [10]. We experiment with fine-tuning the model using EWC together with global BN statistics of O under the following two conditions: fine-tuning only the BN layers, and fine-tuning all layers.

2.3   Datasets and Preprocessing

Institution 1 dataset comprised digital screening mammograms retrospectively collected from 33 institutions through the Digital Mammographic Imaging Screening Trial, the details of which were previously published [18]. This study was approved by the Institutional Review Board (IRB) of the American College of Radiology Imaging Network, by the IRB and the Cancer Therapy Evaluation Program at the National Cancer Institute.

For this Institution 1 dataset, 5 digital mammography systems were used: SenoScan (Fischer Medical), the Computed Radiography System for Mammography (Fuji Medical), the Senographe 2000D (General Electric Medical Systems), the Digital Mammography System (Hologic), and the Selenia Full Field Digital Mammography System (Hologic) [18]. The final Institution 1 patient cohort consisted of 108,230 digital screening images from 21,759 patients.

Institution 2 dataset comprised digital screening mammograms retrospectively collected from a single institution in 2010 following IRB approval. Patients who had prior surgery or implants were excluded. All mammograms were acquired using a Lorad Selenia mammography system (Hologic). The final Institution 2 patient cohort consisted of 8,603 digital screening images from 1,856 patients. All images from Institution 1 and 2 were interpreted by a single radiologist from a pool of radiologists using the ACR BI-RADS breast density lexicon (Category A: fatty, Category B: scattered, Category B: heterogeneously dense, Category D: extremely dense) [13].

The Scanner A (Senographe from Institution 1 dataset, $n = 59411$), Scanner B (Senoscan from Institution 1 dataset, $n = 32928$), Institution 1 ($n = 103890$), Institution 2 ($n = 8603$) patient cohorts were split into training, validation, and testing sets in a 7:2:1 ratio, on a patient level. The intensity of each image was scaled between 0 and 1 before resizing them to $224 \times 224 \times 3$.

2.4   Training and Prediction

[1] For the baseline classification model, a Resnet50 architecture with ImageNet pre-trained weights [5] was used. Cross entropy loss function & the Adam optimizer [9] (lr $= 10^{-6}$, $\beta_1 = 0.9$, $\beta_2 = 0.99$, $\epsilon = 10^{-7}$) were used across models.

---

[1] Code Availability: All code from our method is available at: https://github.com/QTIM-Lab/MedicalDomainExpansion.

Batches were randomly sampled using a fixed batch size of 32 images. Early stopping with a patience of 20 epochs is used to prevent overfitting. Checkpoints were saved after each epoch based on the performance on the validation set & the model with highest validation accuracy was saved as the final model. The training set is augmented in real-time using random flips & rotations (-45–45°).

To combine predictions from all images across all mammography views from a given patient study into a patient-level assessment, the output probabilities for all corresponding images from a patient study are averaged. The averaged probabilities are then used to determine the predicted breast density class. All models are evaluated using Cohen's Kappa scores with linear weighting (κ). For reference, a κ of 0.21–0.40, 0.41–0.60, and 0.61–0.80 represents fair, moderate, and substantial agreement, respectively [20]. A non-parametric two-sided Wilcoxon signed-rank test at a significance level of $p = 0.05$ was used for statistical comparisons of model performance.

3   Results

To fine-tune our deep learning model on T and simultaneously mitigate CF on O, we experiment with two broad sets of approaches:

1. Evaluation with global BN statistics of T (Fig. 1b) Top) vs. O (Fig. 1b) Bottom): both when fine-tuning with only BN layers (i.e. freeze all non-BN layers, allowing only BN parameters to train) & with all layers.
2. Evaluation after incorporating EWC using global BN statistics of O, fine-tuning with only BN layers & with all layers.

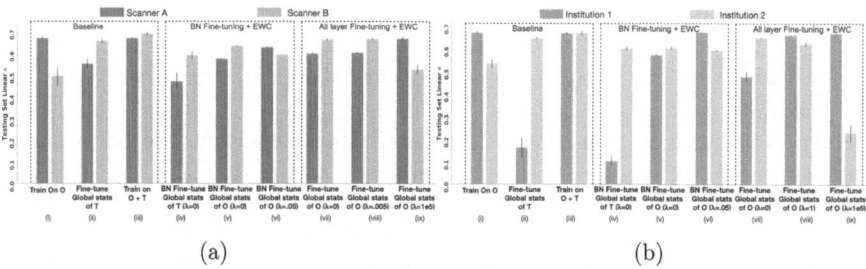

Fig. 2. Results for domain expansions across (a) digital mammography systems (Scanner A to Scanner B); (b) institutions (Institution 1 to Institution 2) The different panels for each domain expansion type correspond to experiments for 1) Baseline models, 2) Only BN-layer fine-tuning with and without EWC [10], and 3) All layer Fine-tuning with and without EWC [10].

3.1 Baseline Experiments

As a baseline, we performed experiments with three approaches: 1) training solely on O, 2) training on O and fine-tuning on T, and 3) training on combined O & T. As shown in Fig. 2a, 2b (Panel "Baseline"), starting from O (Scanner A, Institution 1) to T (Scanner B, Institution 2), we observe that models exclusively trained on O do not generalize well on T (Fig. 2a, 2b i). However, a model originally trained on O, when fine-tuned on T (Fig. 2a, 2b ii), abruptly forgets the information it learned on O ($p < 0.01$). Only when the model is trained collectively O and T (Fig. 2a, 2b iii) does it achieve high performance on both domains (Scanner A κ: 0.67 & Scanner B κ: 0.69, Institution 1 κ: 0.67 & Institution 2 κ: 0.67).

3.2 Global BN Statistics of T vs. O

Fine-Tuning BN Layers: With the intent of aligning the data distribution of T with O, we ran several experiments fine-tuning only the BN layers for T while freezing all the non-BN layers. We start with the baseline model previously trained on O. This model is fine-tuned with two distinct approaches. In the first approach (traditional method as shown in Fig. 1b) Top), this partially frozen model is fine-tuned using the batch statistics of T ($\mu_{batch,T}$ & $\sigma_{batch,T}$) while its global BN statistics ($\mu_{global,T}$ & $\sigma_{global,T}$) are concurrently calculated during the training process. The global BN statistics are then used to evaluate both datasets (Fig. 2a, 2b iv). In the second approach (as shown in Fig. 1b) Bottom), the global BN statistics of O ($\mu_{global,O}$ & $\sigma_{global,O}$) are used for fine-tuning this model on T & again used for evaluation on both the datasets (Fig. 2a, 2b v). From Fig. 2a (Panel"BN Fine-Tuning + EWC"), we observe that when the global BN statistics of Scanner B (T) are used (Fig. 2a iv), the model undergoes a large reduction ($p < 0.01$) in the performance on Scanner A (O). Moreover, the performance on Scanner B (T) is lower ($p < 0.01$) compared to the performance when fine-tuned with all the layers (Fig. 2a ii). In contrast, when the model is fine-tuned using the global BN statistics of Scanner A (O) (Fig. 2a v), we see a recovery ($p < 0.01$) in the performance on Scanner A (O) as well as an increase in the performance on Scanner B (T) from the baseline model before fine-tuning ($p < 0.01$) (Fig. 2a i). Similar results are obtained for Institution 1 (O) and Institution 2 (T). From these results, we can conclude that fine-tuning BN layers using the global BN statistics of O confers a performance advantage over fine-tuning using the global BN statistics of T. Moreover, although the performance on the target domain improves, fine-tuning with BN layers only partially mitigates CF and therefore does not represent successful domain expansion.

Fine-Tuning All Layers: Starting with the same baseline model specification as above, we again experiment with fine-tuning all layers of the model including the BN layers to utilize the full capacity of the model for learning new knowledge from T. Similar to the above setup, this model is fine-tuned for all layers over T and evaluated first using the global BN statistics calculated while training on T (Fig. 2a, 2b ii) and second using the global BN statistics of O (Fig. 2a, 2b vii). When a model trained on O is fine-tuned for all layers on T using the global BN statistics of T (Fig. 2a, 2b ii), we again observe that performance on O degrades ($p < 0.01$ for Scanner A & Institution 1). Evaluating with global BN statistics of O (Fig. 2a, 2b vii) attenuates the performance loss ($p < 0.05$). With this approach, the model can perform well on the target domain (Scanner B & Institution 2) irrespective of the choice of global statistics used.

3.3   Incorporation of EWC

Fine-Tuning BN Layers with EWC: We explore the influence of EWC on the performance of O and T when only the BN layers are fine-tuned using EWC (Fig. 2a, 2bvi). Performance on both datasets is evaluated after fine-tuning on T while varying the importance parameter (λ). After experimenting with a wide range of λ values, we observe that when using the global BN statistics of O we achieve maximum performance on both O & T at $\lambda = 0.05$ (Fig. 3 (a), 3 (b) i).

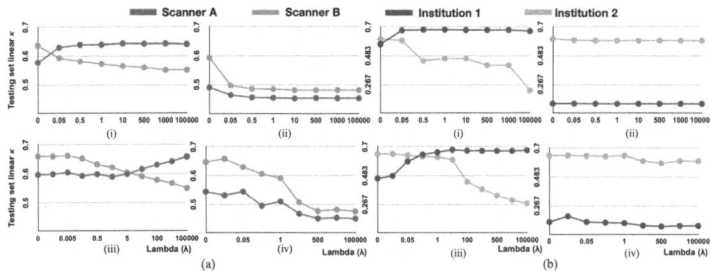

Fig. 3. Model performance while varying λ when fine-tuning on Scanner B or Institution 2 with only BN layers (top row: (i),(ii)) vs. all layers (bottom row: (iii),(iv)) with EWC [10] on (a) Scanner A and Scanner B using global BN statistics of Scanner A (left) vs. Scanner B (right) ; (b) Institution 1 and Institution 2 using global BN statistics of Institution 1 (left) vs. Institution 2 (right)

Fine-Tuning All Layers with EWC: To further mitigate CF on O and achieve peak performance on T, we allow fine-tuning of all layers while also incorporating EWC in the loss function. During fine-tuning with the global BN statistics of O, with increasing λ, performance on O consistently improves (dark red or dark blue in Fig. 3(a), 3(b) iii) respectively. In contrast, for T, we see a degradation in performance with increasing λ (orange or green in Fig. 3(a), 3(b) iii) respectively. When fine-tuning with the global BN statistics of O for very high λ values ($\lambda = 1e + 5$), CF for O is mitigated, with a performance of κ: 0.67 and κ: 0.67 on Scanner A & Institution 1 respectively (Fig. 2a, 2b ix). However, very high λ values prevent the model from learning the features of T,

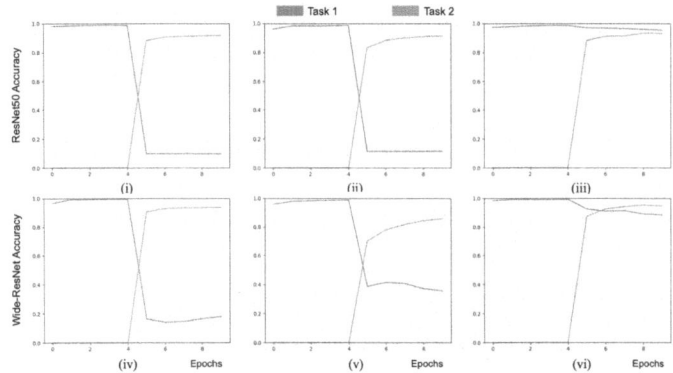

Fig. 4. Performance using Permuted MNIST: ResNet50 architecture (top) (i) standard training (ii) using dropout probability of 0.5 before the last layer (iii) using batch statistics of task 1 while fine-tuning; Wide-ResNet architecture (bottom) (iv) standard training (v) using dropout probability of 0.1 (vi) using batch statistics of task 1 while fine-tuning.

and hence perform poorly on Scanner B (κ: 0.53) & Institution 2 (κ: 0.23) as seen in Fig. 2a, 2b ix.

Interestingly, at the λ which gives the highest performance on both O and T, fine-tuning only BN layers rather than all layers (dark red and dark blue in Fig. 3(a), 3(b) when comparing i) vs. iii) gives a higher performance on O for both scenarios ($p < 0.01$). Conversely, fine-tuning all layers rather than just BN layers (orange or green in Fig. 3(a), 3(b) when comparing i) vs. iii results in higher performance on T ($p < 0.01$). For both BN-only and all Layers fine-tuning, using EWC in conjunction with global BN statistics of T is ineffective for mitigating CF as compared to when using EWC with the global BN statistics of O (Fig 3(a), 3(b)) comparing red or blue curves for left vs. right plots) further substantiating the importance of using global BN statistics of O ($p < .05$). Thus, in summary, domain expansion is optimally performed by using both, the global BN statistic of O (to effectively mitigate CF) and all Layers fine-tuning together with EWC (to attain peak performance on T).

3.4 Analysis Using Permuted MNIST

To understand the advantages of our approach outside medical imaging datasets, we performed analysis using a commonly used permuted MNIST dataset. [16] showcased how using dropout as an implicit gating mechanism in multi-layer perceptron (MLP), they could mitigate CF with permuted MNIST. We attempted to adapt their idea for more modern, deeper network architectures, such as ResNet50. We trained a ResNet50 model on two tasks of permuted MNIST described in their paper. The permuted MNIST dataset is generated by shuffling pixels such that the permutation is the same between images of the same task but is different across the tasks [4]. When a standard Resnet50 architecture is trained on task 1 and fine-tuned on task 2, performance on task 1 drops sharply (Fig. 4 i). In order to adapt [16] to our Resnet50 architecture, we inserted dropout with a probability of 0.5 before the last layer. This led to significant levels of CF with accuracy of 0.12 and 0.91 on validation datasets from task 1 and 2 respectively after training on task 2 (Fig. 4 ii) showing that while their dropout technique works well with a shallow network like MLP, the technique fails for complex model architectures.[2] However, when a ResNet50 architecture was fine-tuned on task 2 using global BN statistics of task 1, CF was significantly mitigated. Performance on task 1 and task 2 was 0.95 and 0.93 respectively (Fig. 4 iii). We found similar results (Fig. 4 iv, v, vi) when experimenting with Wide Resnet architecture [26] with a widening factor of 2 and 50 convolution layers (WRN-50-2), inserting dropouts in intermediate convolutional layers.[3] All in all, our results show that even for commonly used public datasets such as Permuted MNIST, by utilizing our global BN statistics technique, we can easily mitigate CF to a large extent, proving the efficacy of our technique.

4   Discussion

Normalization layers have become ubiquitous in training deep neural networks. BN involves normalizing the output of a layer using the mean & standard deviation of the input batch, effectively allowing training with higher learning rates & improving convergence to a better local optimum [6]. While most existing works focus on the influence of BN when training models on a *single institution dataset*, few have tried to adapt the BN layers for training these models sequentially over multiple datasets with varied scanner/camera types & patient populations. [7] froze the convolutional layers & trained only the BN layers independently for each dataset. Although this approach allows the model to retain performance on the previously trained distribution & task, it restricts the model capacity to learn newer tasks or domains. Moreover, every institution would need to have separate BN parameters, which defeats the purpose of *one model for both the original & target datasets*. In our approach, we allow training of all layers on the target dataset using EWC [10], while simultaneously fixing BN statistics to the global BN statistics of O, thus preventing CF on O. Contrary to [7], our approach provides one universal model for both original & target distributions.

Plasticity is the capacity of a network to adapt to new environments while stability ensures retention of previously learned knowledge. Setting the appropriate trade-off between the stability & plasticity of a neural network is critical,

---

[2] Although we were able to replicate their results (perform domain expansion with minimal CF) using a two-layered MLP (with dropout), we were unable to achieve high performance using a Resnet50 architecture. One possible reason could be that MLP, due to its fully connected layers, is somewhat blind to the permutations and hence does not forget much from task 1 when trained on task 2.

[3] For a dropout probability of 0.10, accuracies for task 1 and 2 were 0.36 and 0.86 respectively. At higher dropout probabilities, the model was unable to converge for task 2.

not only to avoid forgetting but also to learn new tasks quickly. Catastrophic forgetting occurs when a network is overly plastic & not sufficiently stable; *i.e.*, the network can quickly acquire new tasks or modalities but does not retain previously learned tasks and modalities. [16] studied this behavior in the context of dropout layers and showed how dropout can overcome catastrophic forgetting on a previous task when the same model is trained on a new task. Specifically, they showed that multi-layered perceptrons trained with dropout regularization have higher stability & lower variance in the output after ReLU activation function. In other words, activations that are purely active or inactive (values 0 or 1) remain untouched (ensuring stability on previously learned tasks) & only the semi-active neurons are turned off or on, thus promoting plasticity for newer tasks. EWC works on a similar principle, ensuring stability by constraining those weights which are most important for previous learned tasks, while still retaining enough plasticity by allowing other weights to be trained for newer tasks.

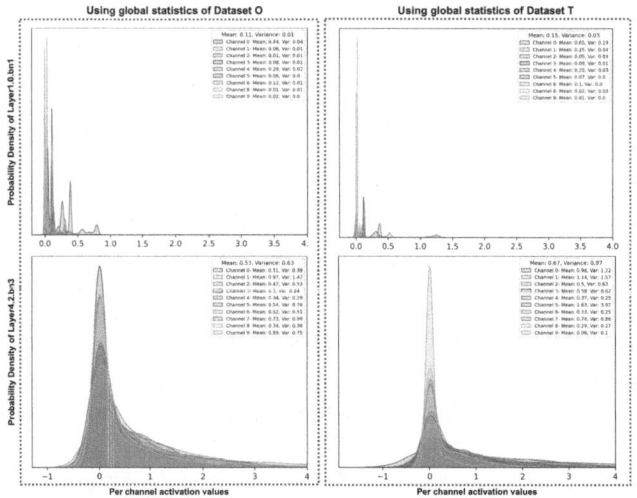

Fig. 5. Probability Density of per-channel activation values computed over a batch from O after fine-tuning on T using global statistics of O (left) & T (right) for the first BN layer in the Bottleneck layer (top) & the last BN layer before the final classification layer (bottom) of the Resnet-50 architecture.

4.1 Analysis of Variance Shifts with Global BN Statistics of T vs. O

To better understand the influence of global BN statistics on the overall stability of a network fine-tuned over T, we plot the distribution of activation values from different BN layer outputs in the network using samples from O (Fig. 5 for Institution 1). Using global BN statistics of O (Fig. 5 left), we empirically see that deeper layers of the network (Fig. 5 top vs. bottom), have a lower activation variance than when using global BN statistics of T (Fig 5 right). Notably, while the mean-variance across different channels for the model fine-tuned using global BN statistics of O increases from 0.01 to 0.63, the model fine-tuned using the global BN statistics of T shows a much greater increase from 0.03 to 0.97. Next, we provide detailed theoretical proof for the variance shifts between global BN statistics of T and O.

From [16], we know that the variance of the penultimate layer's activations directly impacts the stability of the network. In other words, the lower the variance of the activations of the latter layers, the smaller the extent of a model's CF. To illustrate the influence of using the global BN statistics of T and O on the overall stability of a network, we compute the variance of the outputs of the BN layers of a model fully fine-tuned on T (without EWC), given in Eqs. 2 and 3. The variance of a layer's activations for the model trained using global BN statistics of T from Eq. 2 is given by:

$$Var[y_i] = Var\left[\gamma_T \frac{x_i - \mu_{global,T}}{\sqrt{\sigma^2_{global,T} + \epsilon}} + \beta_T\right] \tag{4}$$

Assuming γ_T, β_T, $\mu_{global,T}$ and $\sigma^2_{global,T}$ to be constant at the time of inference and thereafter to understand CF, O is used for inference, $Var[\beta_T] = 0$ and $Var\left[\gamma_T \frac{x_i - \mu_{global,T}}{\sqrt{\sigma^2_{global,T} + \epsilon}}\right] = \frac{\gamma_T^2}{\sigma^2_{global,T} + \epsilon} Var(x_i)$, the final variance of output from BN layer of a model trained using global BN statistics of T is given by:

$$Var[y_i] = C_T \times Var(x_i) \tag{5}$$

where C_T is any constant. Next, the variance for the global BN statistics of O from Eq. 3 is given by:

$$Var[y_i] = Var\left[\gamma_T \frac{x_i - \mu_{global,O}}{\sqrt{\sigma^2_{global,O} + \epsilon}} + \beta_T\right] \tag{6}$$

Again assuming γ_T, β_T, $\mu_{global,O}$ and $\sigma^2_{global,O}$ to be constant at the time of inference and thereafter to understand CF, O is used for inference, $Var[\beta_T] = 0$ and $Var\left[\frac{x_i - \mu_{global,O}}{\sqrt{\sigma^2_{global,O+\epsilon}}}\right] \approx 1$, the final variance of output from BN layer of a model trained using global BN statistics of O is given by:

$$Var[y_i] = \gamma_T^2 = C_T \tag{7}$$

Thus while the variance in the case of the model trained using the global BN statistics of O (Eq. 7) is simply a constant (based on the scaling parameter of the corresponding BN layer), the variance in the case of model trained using the global BN statistics of T (Eq. 5) is a constant times the variance of the input to the BN layer. In other words, though we see some variance shift in both cases, the shift in case of global BN statistics of T (Eq. 5) keeps on increasing as we go deeper into the network (owing to its direct proportionality to input variance). This increase in the variance shift over the layers of a network, ultimately results in extreme distortions in the features computed from the penultimate layers (as seen in Fig. 6) of a model trained using global BN statistics of T when tested on the original dataset O, thus leading to poor stability of the network & ultimately resulting in adverse CF over O.

4.2   UMAP Analysis

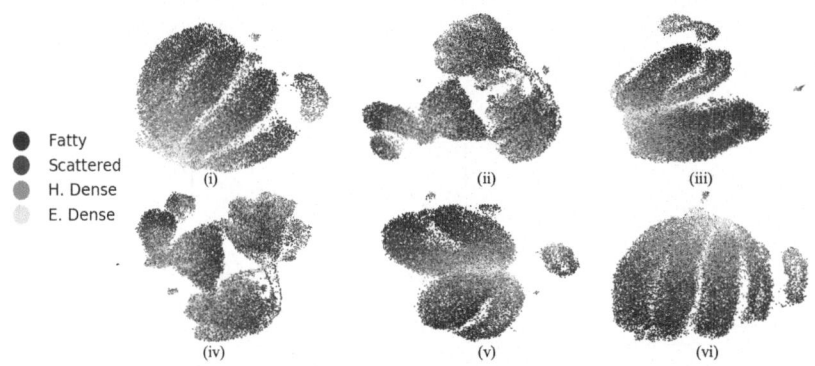

Fig. 6. UMAPs for O when (i) Model is trained just on O; When fine-tuning only BN layers with global BN statistics of (ii) T and (iii) O; Fine-tuning all layers with global BN statistics of (iv) T, (v) O and (vi) O after incorporating EWC

To further investigate the influence of global BN statistics on CF and validate our hypothesis from the previous section, we visualize UMAPs [15] of features computed over O from the various models in our experiments. Figure 6(i) depicts UMAPs of sample features for the model trained originally on O (Institution 1). We observe clear separation for different classes, aligned along a particular direction in low-dimensional space that likely corresponds with high overall performance. In Fig. 6(ii) or 6(iv) (representing models fine-tuned with only BN layers & fine-tuned with all layers respectively by using global BN statistics of T (Institution 2)), we observe that UMAPs of features from O have classes which are highly entangled with each other, which correlate with lower, overall performance on O. However, for Fig. 6(iii) or 6(v) (representing models fine-tuned with only BN layers & fine-tuned with all layers, respectively, by using global BN statistics of O), we observe how different classes which show high separation results in avoiding any CF. This is also observed when fine-tuning all layers after incorporating EWC and using global statistics of O (Fig 6 (vi) shows classes are well segregated & aligned in one direction). Overall, this illustrates how global BN statistics of O prevent large variance shifts in the deeper layers, which may avoid any serious entanglement of features in deeper layers, and ultimately results in higher stability of the network to mitigate CF.

There are several limitations & possible future directions to this work. First, we only evaluate our approach in the context of domain expansion where a model trained on one dataset is fine-tuned on a second dataset. Future work can investigate the scalability of our approach when domain transfer is performed across a larger number of domains or datasets. Second, both clinical questions investigated in this study were classification tasks. Additional studies evaluating the utility of our approach for other clinical tasks such as detection & segmentation are warranted. Finally, federated learning (FL) has received considerable attention recently, primarily for training models over multi-institutional medical datasets, which are difficult to share due to both privacy and legal issues [21, 22]. FL presents an alternative to synchronous fine-tuning by training on multiple datasets simultaneously. However, data heterogeneity remains an optimization challenge for FL. Comparison of our approach with FL under real-world settings can be a topic of future investigation.

References

1. Chang, K., et al.: Multi-institutional assessment and crowdsourcing evaluation of deep learning for automated classification of breast density. J. Am. Coll. Radiol. **17**(12), 1653–1662 (2020)
2. Chen, T., Kornblith, S., Norouzi, M., Hinton, G.: A simple framework for contrastive learning of visual representations. arXiv preprint arXiv:2002.05709 (2020)
3. Esteva, A., Kuprel, B., Novoa, R.A., Ko, J., Swetter, S.M., Blau, H.M., Thrun, S.: Dermatologist-level classification of skin cancer with deep neural networks. Nature **542**(7639), 115–118 (2017)
4. Goodfellow, I.J., Mirza, M., Xiao, D., Courville, A., Bengio, Y.: An empirical investigation of catastrophic forgetting in gradient-based neural networks. arXiv preprint arXiv:1312.6211 (2013)
5. He, K., Zhang, X., Ren, S., Sun, J.: Deep residual learning for image recognition. In: Proceedings of the IEEE Conference on Computer Vision and Pattern Recognition, pp. 770–778 (2016)
6. Ioffe, S., Szegedy, C.: Batch normalization: accelerating deep network training by reducing internal covariate shift. arXiv preprint arXiv:1502.03167 (2015)
7. Karani, N., Chaitanya, K., Baumgartner, C., Konukoglu, E.: A lifelong learning approach to brain MR segmentation across scanners and protocols. In: International Conference on Medical Image Computing and Computer-Assisted Intervention, pp. 476–484. Springer (2018). https://doi.org/10.1007/978-3-030-00928-1_54
8. Keavey, E., Phelan, N., O'Connell, A., Flanagan, F., O'Doherty, A., Larke, A., Connors, A.: Comparison of the clinical performance of three digital mammography systems in a breast cancer screening programme. Br. J. Radiol. **85**(1016), 1123–1127 (2012)
9. Kingma, D.P., Ba, J.: Adam: a method for stochastic optimization. arXiv preprint arXiv:1412.6980 (2014)
10. Kirkpatrick, J., Pascanu, R., Rabinowitz, N., Veness, J., Desjardins, G., Rusu, A.A., Milan, K., Quan, J., Ramalho, T., Grabska-Barwinska, A., et al.: Overcoming catastrophic forgetting in neural networks. Proc. Natl. Acad. Sci. **114**(13), 3521–3526 (2017)

11. Li, M.D., et al.: Siamese neural networks for continuous disease severity evaluation and change detection in medical imaging. NPJ digital medicine **3**(1), 1–9 (2020)
12. Li, Y., Wang, N., Shi, J., Liu, J., Hou, X.: Revisiting batch normalization for practical domain adaptation. arXiv preprint arXiv:1603.04779 (2016)
13. Liberman, L., Abramson, A.F., Squires, F.B., Glassman, J., Morris, E., Dershaw, D.D.: The breast imaging reporting and data system: positive predictive value of mammographic features and final assessment categories. AJR Am. J. Roentgenol. **171**(1), 35–40 (1998)
14. Ly, A., Marsman, M., Verhagen, J., Grasman, R., Wagenmakers, E.J.: A tutorial on fisher information (2017)
15. McInnes, L., Healy, J., Melville, J.: UMAP: uniform manifold approximation and projection for dimension reduction. arXiv preprint arXiv:1802.03426 (2018)
16. Mirzadeh, S.I., Farajtabar, M., Ghasemzadeh, H.: Dropout as an implicit gating mechanism for continual learning. arXiv preprint arXiv:2004.11545 (2020)
17. Mohamed, A., Berg, W., Peng, H., Luo, Y., Jankowitz, R., Wuand, S.: A deep learning method for classifying mammographic breast density categories. Med. Phys. **45**(1), 314–321 (2017)
18. Pisano, E.D., Gatsonis, C., Hendrick, E., Yaffe, M., Baum, J.K., Acharyya, S., Conant, E.F., Fajardo, L.L., Bassett, L., D'Orsi, C., et al.: Diagnostic performance of digital versus film mammography for breast-cancer screening. N. Engl. J. Med. **353**(17), 1773–1783 (2005)
19. Razzaghi, H., Troester, M.A., Gierach, G.L., Olshan, A.F., Yankaskas, B.C., Millikan, R.C.: Mammographic density and breast cancer risk in white and african american women. Breast Cancer Res. Treat. **135**(2), 571–580 (2012)
20. Richard, L., Gary, K.: The measurement of observer agreement for categorical data. arxiv e-prints, page. Biometrics **33**, 159–174 (1977)
21. Roth, H.R., et al.: Federated learning for breast density classification: a real-world implementation. LNCS, pp. 181–191 (2020). https://doi.org/10.1007/978-3-030-60548-3_18, http://dx.doi.org/10.1007/978-3-030-60548-3_18
22. Sheller, M.J., Edwards, B., Reina, G.A., Martin, J., Pati, S., Kotrotsou, A., Milchenko, M., Xu, W., Marcus, D., Colen, R.R., et al.: Federated learning in medicine: facilitating multi-institutional collaborations without sharing patient data. Sci. Rep. **10**(1), 1–12 (2020)
23. Sprague, B.L., Conant, E.F., Onega, T., Garcia, M.P., Beaber, E.F., Herschorn, S.D., Lehman, C.D., Tosteson, A.N., Lacson, R., Schnall, M.D., et al.: Variation in mammographic breast density assessments among radiologists in clinical practice: a multicenter observational study. Ann. Intern. Med. **165**(7), 457–464 (2016)
24. Sprague, B.L., et al.: Variation in mammographic breast density assessments among radiologists in clinical practice: a multicenter observational study. Ann. Intern. Med. **165**(7), 457–464 (2016)
25. Yu, T., Bagdasaryan, E., Shmatikov, V.: Salvaging federated learning by local adaptation. arXiv preprint arXiv:2002.04758 (2020)
26. Zagoruyko, S., Komodakis, N.: Wide residual networks. arXiv preprint arXiv:1605.07146 (2016)
27. Zech, J.R., Badgeley, M.A., Liu, M., Costa, A.B., Titano, J.J., Oermann, E.K.: Variable generalization performance of a deep learning model to detect pneumonia in chest radiographs: a cross-sectional study. PLoS Med. **15**(11), e1002683 (2018)
28. Zeng, G., Chen, Y., Cui, B., Yu, S.: Continual learning of context-dependent processing in neural networks. Nature Machine Intelligence **1**(8), 364–372 (2019)

Distribution-Aware Replay for Continual MRI Segmentation

Nick Lemke[1]([✉]), Camila González[2], Anirban Mukhopadhyay[1], and Martin Mundt[1,3]

[1] Technical University of Darmstadt, Darmstadt, Germany
nick.lemke@gris.informatik.tu-darmstadt.de
[2] Stanford University, Stanford, USA
[3] The Hessian Center for Artificial Intelligence: hessian.AI, Darmstadt, Germany

Abstract. Medical image distributions shift constantly due to changes in patient population and discrepancies in image acquisition. These distribution changes result in performance deterioration; deterioration that continual learning aims to alleviate. However, only adaptation with data rehearsal strategies yields practically desirable performance for medical image segmentation. Such rehearsal violates patient privacy and, as most continual learning approaches, overlooks unexpected changes from out-of-distribution instances. To transcend both of these challenges, we introduce a distribution-aware replay strategy that mitigates forgetting through auto-encoding of features, while simultaneously leveraging the learned distribution of features to detect model failure. We provide empirical corroboration on hippocampus and prostate MRI segmentation. To ensure reproducibility, we make our code available at https://github.com/MECLabTUDA/Lifelong-nnUNet/tree/cl_vae.

Keywords: Continual Learning · Out-of-Distribution Detection

1 Introduction

Deep learning approaches are largely regarded as successful in static biomedical image segmentation settings [14]. Yet, medical data may shift according to changes in the patient population, vary according to disease-related factors, or be subject to differences resulting from nuances in image acquisition parameters [33]. Since medical image segmentation models are typically trained on small datasets (judged by deep learning standards), they tend to not generalize well to such *shifted distributions* [6]. Ideally, a learner should be able to expand its knowledge by training on new samples from the prospectively shifted or later recorded distributions. As do medical experts, our artificial system should *learn continually* [25]. In order to enable the latter it is required to overcome a phenomenon understood as *catastrophic forgetting* [24,29], or more intuitively, to avoid new information from greedily overwriting existing knowledge.

F. Proietto Salanitri et al. (Eds.): PILM 2024/AIPAD 2024, LNCS 15197, pp. 73–85, 2025.
https://doi.org/10.1007/978-3-031-73483-0_7

However, in medical imaging, continual learning algorithms are so far not the remedy that was promised. Among the conceptual pillars of proposed algorithms [26], rehearsal of data subsets [31] performs by far the best, yet directly violates inherent (medical) *privacy* regulations [35]. Generative replay [34] aims at capturing the distributions encountered during training, and including synthesized data in future training tasks. However, compared to distributions of natural images, those of MRIs are much more difficult to grasp as MRIs are more complex and more high-dimensional. Alternative methods that instead rely on constraining model parameters, so-called regularization approaches [18,37], have in turn been shown to perform poorly on medical data [8]. In fact, this failure mode of forgetting due to having no access to past data is further exacerbated by an often overlooked additional phenomenon - the *silent failure* of models. They not only suffer from expected forgetting of past experiences, but also produce overly confident false predictions whenever unexpected data is encountered [2]. Again ideally, the learner should be able to detect and outright reject these *out-of-distribution* (OoD) examples. Unfortunately, the latter is substantially challenged by the reality that predominant segmentation models like UNet [14,32] lack a notion of the learned distribution. Existing OoD detection algorithms thus often assume a-priori knowledge of the anticipated OoD samples [5,20] or hope that expensive uncertainty approximations capture the examples [15,22]. On the contrary, generative models [17] (that explicitly learn the distribution) are notoriously hard to train for discriminatory tasks.

In this work, we simultaneously address the challenge of avoiding forgetting without direct violations of privacy in continual learning and overcome silent prediction failures by rejecting OoD instances. To this end, we leverage prior insights on theoretically grounded two-stage modeling [4,13], where a second generative model encodes the distribution of our primary discriminative model, without interfering in the latter's learning or inference processes. Specifically, we propose a second-stage conditional variational autoencoder (VAE) [17] to model the low-dimensional distribution of a UNet's latent features. With the feature distribution captured by the VAE we can then make rigorous decisions to assess whether a new subject is outside the known distribution and conversely employ a pseudo-rehearsal setup to replay features of past subjects to avoid forgetting when adapting the model continually. We evaluate our setup on domain incremental MRI segmentation tasks of the hippocampus and the prostate and further assess the OoD detection capabilities on augmented datasets.

2 Methodology

The UNet architecture [32] is well-known for its extraordinary performance in medical image segmentation [14]. How do we leverage this architecture and retain its efficacy while overcoming its inherent forgetful nature and its silent failure modes? To achieve symbiosis between these desiderata we leverage recent theoretical insights [4], proving that a second VAE can correctly model an initial VAE's learned distribution as an isotropic Gaussian distribution as a consequence of the known hidden dimensionality of the first model. This in turn

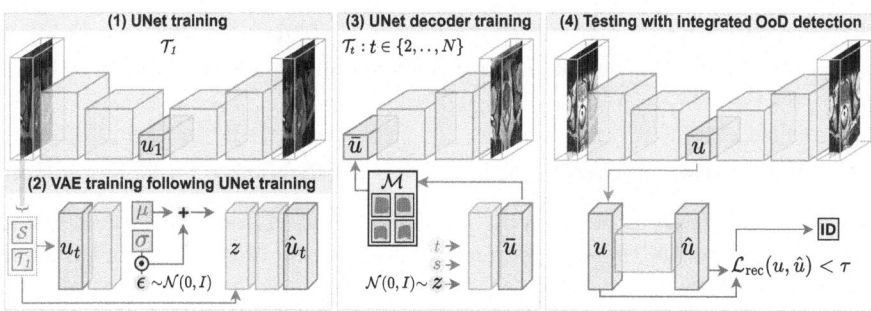

Fig. 1. (1) The UNet is trained on the first task T_1. (2) The VAE is trained on features u_1 with slice and task conditioning. (3) A set of features $\bar{u}_{i<t}$ are synthesized, pseudo-labeled and placed in memory $\mathcal{M}$. The UNet decoder is then trained on $\mathcal{M}$ and the new data of task T_t. (4) During inference, the reconstruction loss between u and $\hat{u}$ is used to classify whether the MRI is ID or OoD.

allows to replay the learned distribution in continual learning [13]. As we will proceed to elaborate, placing such a VAE meta-model on top of a medical UNet will now allow us to i) model and rehearse the feature distribution of a UNet without interfering in its learning process, ii) strategically condition the VAE on observed tasks and volumetric slicing of the medical data, iii) leverage the represented feature distribution to reject OoD examples to avoid silent model failure. Figure 1 shows a schematic representation of the proposed architecture.

2.1 A Two-Stage Architecture for Continual Medical Segmentation

Consider a UNet composed of several blocks of convolutions to downsample the data and then recombine the representation to produce a segmentation map. Conceptually, a UNet is comprised of an encoder, encoding the features of the data into a latent code u, followed by a decoder, decoding the code into the desired output. However, as the model is trained in a supervised discriminative fashion, we unfortunately do not know the form of the distribution of u. We overcome this hurdle by capturing $p(u)$ through a separate VAE. The goal of this model is to learn an approximate posterior $q(z|u)$ through variational inference, where z is a second set of latent factors which we optimize to follow a pre-defined prior $p(z)$. This prior is an easy-to-sample Normal distribution. The key is that the latent code z has the same dimensionality as u. Thus, we can encourage the VAE to learn a lossless mapping from our UNet's unknown feature distribution $p(u)$ to our prior with the aid of a decoder that models the likelihood $p(u|z)$. We can then train the VAE with an evidence lower bound: $\log p(u) \geq \mathbb{E}_{z \sim q(z|u)}\left[\log p(u|z)\right] - \text{KL}\left[q(z|u)\|p(z)\right]$. Here KL denotes the Kullback-Leibler divergence. The UNet training is shown in Fig. 1 (1), followed by the VAE training after each UNet update step in Fig. 1 (2). On arrival of a new task $T_{t>2}$ a buffer $\mathcal{M}$ of pseudo-samples is synthesized by the VAE posterior and pseudo-labeled by the latest UNet decoder. The pseudo-elements and the data from the new task are used to update the UNet decoder as shown in Fig. 1 (3). At the end of the training loop, the VAE is updated using the same memory buffer and the new data (Fig. 1 (2)).

2.2   Distribution-Aware Pseudo-Replay with Native OoD Detection

Intuitively, our UNet first trains on a task $\mathcal{T}_1$ (Fig. 1 (1)) and subsequently the VAE learns to model the encoded feature distribution (Fig. 1 (2)). In principle this already allows us to 1) assess whether new samples are dissimilar to already observed ones, 2) rehearse previous experience by generating pseudo-data [30]. However, to adequately maintain knowledge of each task we have observed in continual learning, we further condition our VAE on the task identity t, i.e. t is appended to the VAE input u and the latent variable z. As the learned task embedding encodes the unique properties of each domain, the VAE remains fixed in size as more distributions are captured.

This conditioned VAE entails multiple advantages. For the above first ability, OoD detection, it enables us to use the VAE's predicted log-likelihood (the reconstruction loss) to decide whether a new sample during UNet inference is dissimilar to any previous tasks' distribution. Once the VAE observes more than one task, we consider the lowest reconstruction error obtained with each previous task identity t. Specifically, we classify samples with a reconstruction error below a threshold τ as in-distribution (ID), which we calibrate on the 95% true positive rate on the validation set (Fig. 1 (4)). Importantly, such an OoD detection procedure does not interfere with the UNet's semantic segmentation prediction, maintaining it's well-known precision and merely augmenting it with an OoD score to inform the user of (un-)trustworthy predictions.

For the second ability, mitigation of forgetting, we use the conditioned VAE to generate pseudo-features $\bar{u}_{i<t}$ for past experiences in the former sequence of tasks $\mathcal{T}_1, \mathcal{T}_2, \ldots$. Here, the task conditioning ensures that we can synthesize a balanced memory $\mathcal{M}$. Specifically, as we progress through tasks the MRIs are first encoded to features u using the UNet encoder, on which the VAE trains with the additional conditioning. To avoid forgetting of these tasks when proceeding to a new task $\mathcal{T}_{t+1}$, we then fill a memory of synthesized examples by: 1) sampling z for each respective task $z \sim p(z|t)$ from the Normal distribution in our VAE, 2) using its decoder to map this random value to a UNet's pseudo-feature $\bar{u}$ that is alike previous experience, and 3) inferring the pseudo-feature's label with the UNet decoder (Fig. 1 (3)). To ensure that the distribution of features does not change as we continue training the decoder, we freeze the UNet's encoder after the first task. Finally, after each task's training, the encoded features of the UNet are then deleted, and the current memory is flushed to reduce the memory footprint and ensure adherence to privacy considerations.

2.3   MRI Advantages Through VAE Double Conditioning

Following theory [4], the distribution is only correctly learned by the VAE if its latent dimension matches the UNet encoder's feature dimensionality:

$\dim(z) == \dim(u)$. Though already low-dimensional, our 3D UNet still has a spatial resolution of $5 \times 7 \times 5$ with 256 channels. This results in a latent space of size $5 \times 7 \times 5 \times 256 = 44,800$, which remains cumbersome. To make our final model computationally feasible, we restrict the UNet to be two-dimensional by segmenting slice-wise along the lowest resolution, reducing the dimension by a factor of 5 to 8,960. The two-dimensional UNet is thus applied to slices of the 3D image volume and the smaller latent space is well learnable by the VAE. However, we now expect large differences in the features between slices at different locations in the volume. To ensure that this choice does not become detrimental, we introduce a final conditioning into the VAE: a further slice index s to indicate the position of the slice within the volume. We refer to this doubly-conditioning architecture as **ccVAE** and show its empirical superiority in the following.

3   Experimental Setup

Data: Following previous work on medical continual segmentation [9,10,27,28], we evaluate on the tasks of segmenting the prostate and hippocampus in, respectively, T2-weighted and T1-weighted MRIs. The **hippocampus** data consists of three datasets: *Multi-contrast submillimetric 3 T hippocampal subfield segmentation (Dryad)* [19], *Harmonized Hippocampal Protocol dataset (HarP)* [36] and the hippocampus data released for the *Medical Segmentation Decathlon (Decath-Hip)* [1]. We train in the order *DecathHip→Dryad* following the setup in previous works [8]. We preserve *HarP* for OoD testing. The sets contain 260, 50, and 270 samples, respectively. The **prostate** data originates from five institutions using different devices and acquisition parameters [23]. We train in the order *BIDMC→I2VCB→HK→UCL*, creating a challenging setting by starting with the smallest dataset and alternating between datasets with and without an endorectal coil. The segmentation mask encompasses the central gland and peripheral area. We likewise use the final dataset, *RUNMC* for OoD evaluation. Each dataset contains 12 to 30 samples and is randomly divided into 20% testing, 56% training, and 24% validation. A qualitative comparison of the data used can be found in Fig. 2. We also utilize synthetic OoD data. Here, we augment the test sets with common MRI artifacts (random bias field, spiking, or ghosting) doubling their size. A few examples of augmented MRIs are depicted in Fig. 3.

Architectures and Training: We use the state-of-the-art nnUNet framework [14], which automatically configures UNet parameters based on data characteristics. Our VAE consists of 8 linear layers with batch norm and leaky ReLU, and is trained for 5000 epochs. We use the Adam optimizer [16], an initial learning rate of $1e-3$, and exponential learning rate decay with a rate of $9.9e-1$. For generated features there are no activations in the UNet encoder, so we discard the skip connections. We run our experiments on two Nvidia A40 GPUs.

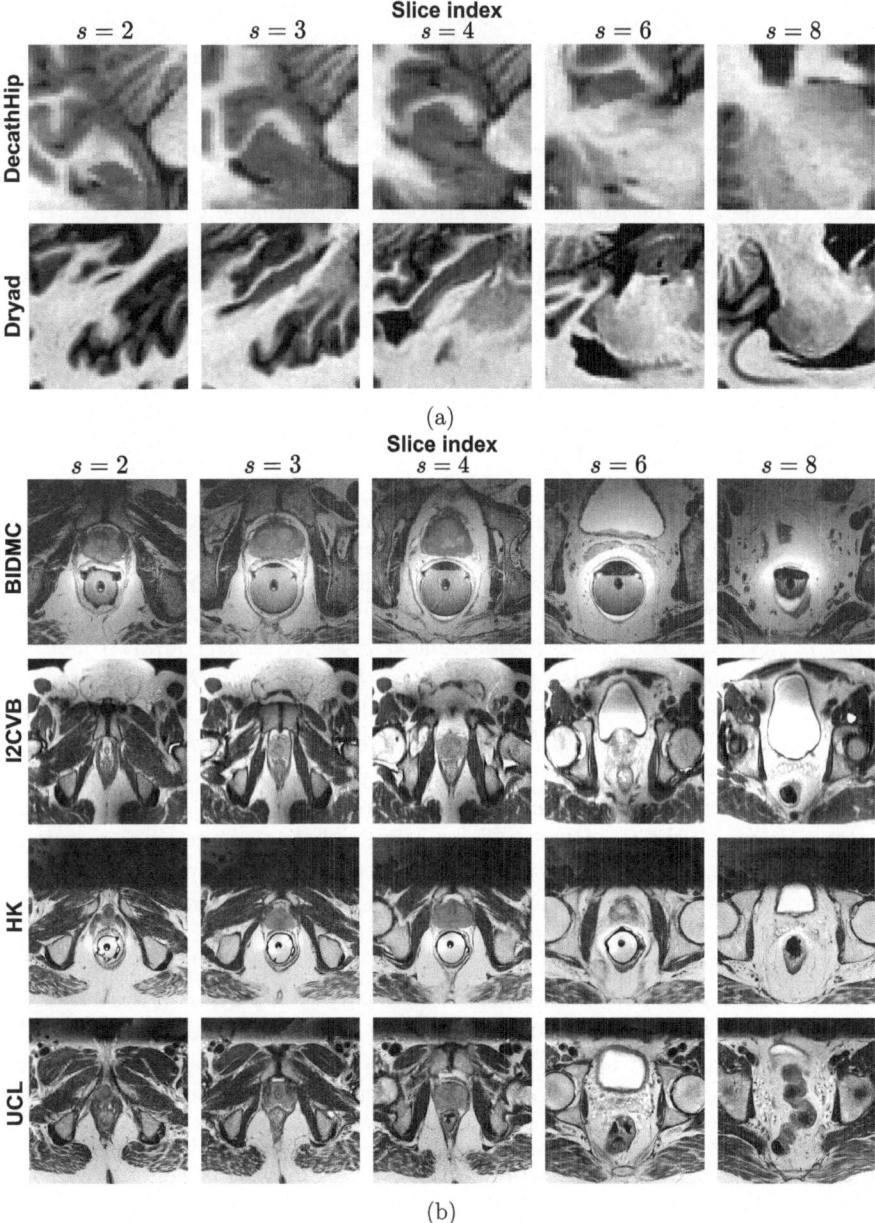

Fig. 2. Representative slices s of MRI scans from each (a) hippocampus and (b) prostate dataset. The red areas depict the ground truth segmentation masks. (Color figure online)

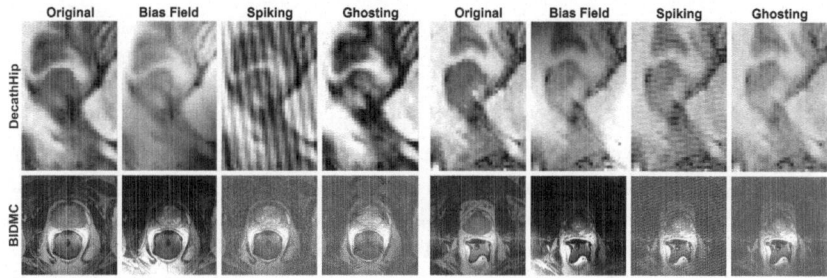

Fig. 3. Augmentations applied to the hippocampus (top row) and prostate (bottom row) datasets to create challenging OoD scenarios.

Baselines: We compare ccVAE to regular sequential (*Seq.*) training and several continual learning methods with comparable privacy preservation. Elastic weight consolidation (*EWC*) [18] penalizes the deviation of parameters deemed to be significant for past tasks. Modeling the background (*MiB*) [3] is tailored specifically for semantic segmentation and uses an unbiased distillation loss that penalizes a shift in the foreground classes. For OoD detection during testing, we use the maximum softmax probability (*SM*) [12]. We also compare to maintaining a pool of models trained at different stages (*MPool*) [9] and using Segmentation Distortion (*SD*) [21] for OoD detection, which similarly to our approach uses an autoencoder for reconstructing features of a pre-trained UNet. During inference, the UNet corresponding to the autoencoder with the lowest SD is chosen for segmentation. Finally, we do an ablation of ccVAE by using only conditioning on the task (*cVAE*) and detecting OoD samples based on the Mahalanobis distance in the feature space (*Mah*) [7] instead of the reconstruction error.

Metrics: For evaluating the segmentation performance of continually trained models, we compute the Dice score for the samples classified as in-distribution. We also report the expected calibration error (ECE) [11] after normalization, as well as the backward (BWT) and forward (FWT) transferability [10].

4 Results

We first evaluate ccVAE in a challenging setting with abrupt shifts in the data distribution during continual training. We further introduce OoD data during testing, first in the form of an unseen dataset and later by adding image artifacts.

Continual Learning Under Dataset Shift: Figure 4 illustrates the performance of ccVAE alongside existing methods in a continual learning context, where new tasks are introduced at 250 epoch intervals. The y-axis depicts the mean Dice for test images from all tasks that are considered ID. After the shift in the hippocampus data, only ccVAE learns to adapt while still producing high-quality segmentations for the older distribution, consequently maintaining robust performance across the trajectory. The expansion-based pooling baseline with segmentation distortion also remains mostly unaffected by the shift but is outperformed by ccVAE. Continual segmentation of the prostate proves more challenging. There is an abrupt fall in segmentation quality after the second task is introduced, likely due to the small size of the database (7 to 11 samples per task) that makes generalization more challenging. As ccVAE recognizes samples from more than the present task as ID and attempts to segment them, we see the performance on T_1 deteriorate. However, from that point on, ccVAE remains stable while other methods display noticeable volatility in segmentation performance.

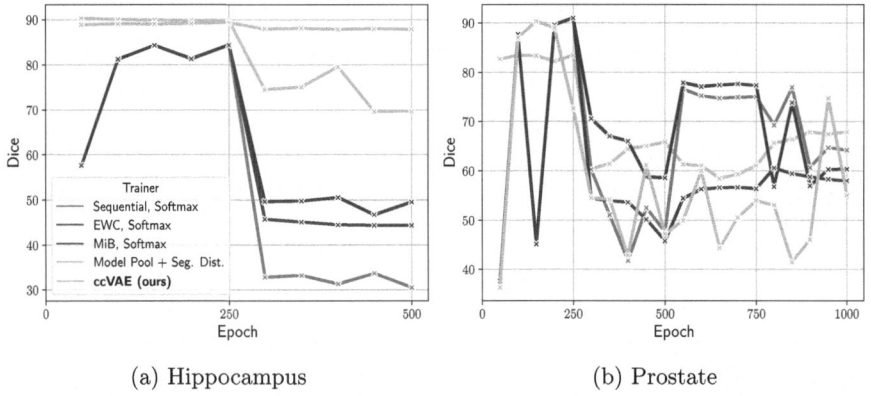

(a) Hippocampus   (b) Prostate

Fig. 4. Test Dice (↑) during the learning trajectory for (a) hippocampus and (b) prostate. New tasks are introduced at 250 epoch intervals. ccVAE (yellow) maintains the most stable segmentation performance throughout the trajectory. (Color figure online)

Table 1. Mean Dice, backward transfer (BWT) and forward transfer (FWT) of the model for all test samples after training on the hippocampus and prostate sequences, respectively. ccVAE achieves the best segmentation performance, with little forgetting and robust knowledge accumulation.

Anatomy/	Hippocampus			Prostate		
Method	Dice ↑	BWT ↑	FWT ↑	Dice ↑	BWT ↑	FWT ↑
Sequential	20.1±32.1	−83.2±8.2	0.0±0.0	54.7±30.9	−43.3±29.6	0.0±0.0
EWC [18]	77.5±28.0	0.0±0.2	−77.3±6.3	53.5±28.8	2.1±8.6	−47.0±28.4
MiB [3]	60.6±16.9	−34.9±10.7	−1.1±0.8	53.3±32.0	−45.6±27.9	0.4±3.4
MPool [9]	72.8±33.0	−13.2±31.5	−37.5±36.3	54.8±35.0	0.9±41.6	−44.1±35.6
ccVAE (ours)	**87.8±4.5**	−1.3±4.8	−3.9±2.0	**64.5±9.1**	−11.4±10.1	−17.0±7.7

Table 1 reports the average Dice, BWT and FWT after the entire training sequence, regardless of whether samples are considered ID or OoD. Sequential training and MiB suffer from substantial forgetting, shown by a large negative BWT and overall lower Dice scores. The expansion-based MPool successfully prevents forgetting, yet at the cost of a loss in plasticity as most members from the model pool do not acquire knowledge from the latter training stages.

Navigating Dataset Shift and Image Artifacts: We now increase the difficulty of the training conditions further by augmenting the test images with synthetically generated MRI artifacts. Table 2 shows the Dice of all images deemed to be ID, alongside the expected calibration error calculated on all test samples. We report the results after each training stage. ccVAE consistently performs well in early stages, showing its ability to identify cases that it can segment successfully. All methods struggle after training with *HK* (column 5), which proves particularly challenging. Here, sequential and MiB training perform well in a trade-off that only considers images from the latest task as ID, disregarding the earlier tasks. As they are both highly plastic methods, they quickly adapt to this new task. ccVAE, on the other hand, considers most images following distributions seen in the past as ID. This demonstrates that despite having some protection against forgetting in the form of generated pseudo-samples, a highly shifted dataset in the sequence will damage the segmentation ability. Still, performance of ccVAE across the trajectory and within each evaluation round remains stable, as corroborated by the consistently low standard deviation in ccVAE predictions.

Table 2. Dice for subjects classified as ID and expected calibration error (**ECE**) after each training stage for all the test data, including cases from each task as well as scans augmented with MRI artifacts. Except for HK, where Seq. SM and MPool trade-off performance, ccVAE demonstrates superior stable performance.

Training stage/ Method	DecathHip Dice ↑	**E** ↓	Dryad Dice ↑	**E** ↓	BIDMC Dice ↑	**E** ↓	I2CVB Dice ↑	**E** ↓	HK Dice ↑	**E** ↓	UCL Dice ↑	**E** ↓
Seq., SM [12]	63.4±39	51.1	19.4±31	48.3	50.5±40	39.8	38.8±36	40.3	71.0±16	26.7	58.9±28	16.7
EWC [18], SM [12]	63.4±39	51.1	32.6±38	49.6	50.5±40	39.8	37.3±32	34.2	46.2±27	30.2	48.2±26	25.3
MiB [3], SM [12]	63.4±39	51.1	26.5±31	45.3	50.5±40	39.8	44.3±30	20.6	70.7±16	21.8	48.5±33	31.8
MPool [9], SD [21]	82.4±24	48.3	47.8±40	42.4	47.2±42	37.2	37.6±34	43.4	46.4±34	37.2	41.4±36	34.4
ccVAE (ours)	89.3± 3	7.8	83.2±14	4.7	75.6±11	14.8	56.7±17	21.5	49.4±21	27.8	58.8±15	32.3

Qualitative Evaluation: Figure 5 illustrates four exemplary prostate segmentations produced by ccVAE. The first and second images are ID MRIs that are correctly classified as such and segmented well. The third is an OoD MRI that is segmented poorly but rejected by the OoD detection mechanism. The fourth MRI is augmented with a ghosting artifact and not detected.

82   N. Lemke et al.

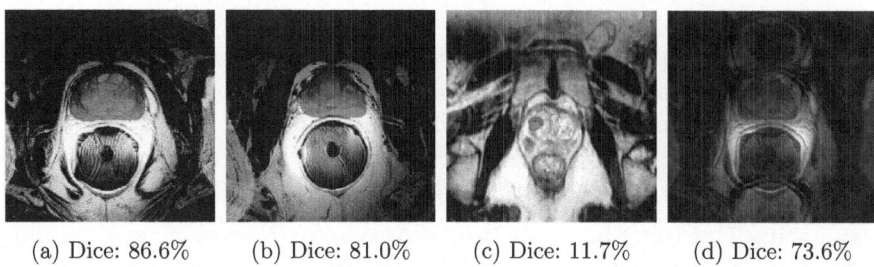

(a) Dice: 86.6%   (b) Dice: 81.0%   (c) Dice: 11.7%   (d) Dice: 73.6%

Fig. 5. Four segmentations produced by the model trained on the first prostate dataset. Images (a) and (b) are correctly considered ID and segmented correctly. (c) is correctly considered OoD, but (d) is misclassified.

Ablation Study: In Table 3 we ablate ccVAE in the simpler scenario without artifact augmentations to corroborate that all elements of our approach are needed. We compare the proposed ccVAE, which detects OoD samples based on the reconstruction error, to estimating the uncertainty from the Mahalanobis distance to the prior distribution $p(z)$ (*ccVAE, Mah.*). We also evaluate a version of the VAE that is only conditioned on the task (*ccVAE, Rec.*). Alongside these ablations, we include the per-stage results of the model pool with segmentation distortion baseline (*MPool SD*), which is closest in performance to ccVAE in Fig. 4. In most stages, the full ccVAE is necessary to obtain the highest Dice and the first or second-lowest ECE. The OoD detection strategy based on the Mahalanobis distance fails to calibrate the model in early training, resulting in high ECEs and low Dice scores.

Table 3. Ablation study comparing ccVAE to different versions of our method and the best baseline from Fig. 4. Both conditioning and basing OoD detection on VAE reconstructions consistently contribute to performance.

Training stage/ Method	DecathHip Dice ↑	E ↓	Dryad Dice ↑	E ↓	BIDMC Dice ↑	E ↓	I2CVB Dice ↑	E ↓	HK Dice ↑	E ↓	UCL Dice ↑	E ↓
MPool [9], SD [21]	89.8± 3	33.4	69.7±35	20.1	72.3±34	30.3	48.6±34	35.1	55.1±31	31.8	55.9±34	30.2
ccVAE, Mah. [7]	89.0± 3	13.2	61.2±33	24.4	39.1±30	29.0	60.5±13	34.7	60.4±18	34.2	67.9±10	22.6
cVAE, Rec.	89.3± 3	3.8	87.6± 4	16.8	83.4± 2	24.4	64.7± 9	19.4	65.4±12	17.3	65.4±10	28.6
ccVAE	89.4± 3	4.7	87.9± 5	14.5	83.4± 2	25.5	66.2± 9	27.2	60.0±19	35.5	67.9±10	37.8

Analysis of Generated Features: Finally, in Fig. 6 we qualitatively support our quantitative findings by visualizing segmentation masks of the train set and similar segmentation masks of the ccVAE's generated features included in pseudo-rehearsal training. The generated features are semantically coherent, cover multiple volume segments and successfully capture geometric diversity.

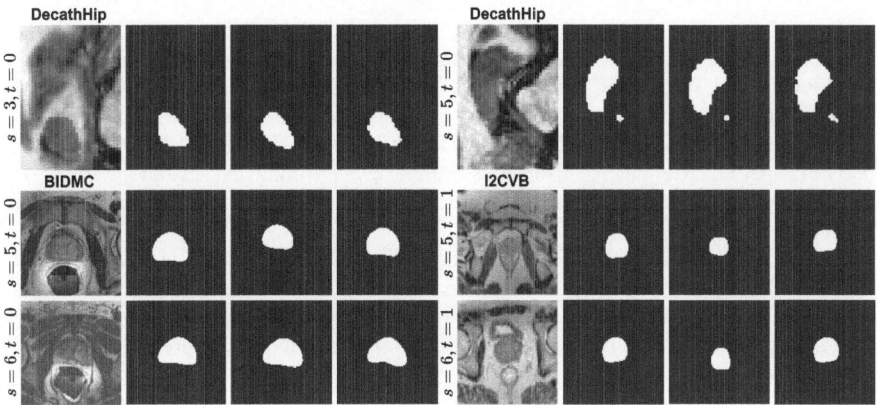

Fig. 6. Ground truth segmentation masks from the original tasks and segmentation masks from generated features using different slice and task indices.

5   Conclusion

Aiming to translate the success of medical image segmentation to more realistic dynamic settings, where there are abrupt shifts in the training distribution and the model encounters low-quality images during testing, we propose ccVAE. Our method augments UNet segmentation models with a small VAE that maps features into a standard normal distribution without reducing dimensionality. In turn, this allows to generate features similar to those seen in previous tasks, preventing forgetting without compromising patient privacy, and enabling principled OoD detection. ccVAE, therefore, jointly addresses the two main factors causing unexpected performance deterioration in dynamic clinical environments.

References

1. Antonelli, M., et al.: The medical segmentation decathlon. Nat. Commun. **13**(1), 4128 (2022)
2. Boult, T.E., Cruz, S., Dhamija, A.R., Gunther, M., Henrydoss, J., Scheirer, W.J.: Learning and the unknown : surveying steps toward open world recognition. In: The AAAI Conference on Artificial Intelligence (2019)
3. Cermelli, F., Mancini, M., Bulo, S.R., Ricci, E., Caputo, B.: Modeling the background for incremental learning in semantic segmentation. In: Proceedings of the IEEE/CVF Conference on Computer Vision and Pattern Recognition (2020)
4. Dai, B., Wipf, D.: Diagnosing and enhancing VAE models. In: International Conference on Learning Representations (2018)
5. Dhamija, A.R., Günther, M., Boult, T.: Reducing network agnostophobia. In: Advances in Neural Information Processing Systems, vol. 31 (2018)
6. Geirhos, R., et al.: Shortcut learning in deep neural networks. Nat. Mach. Intell. **2**(11), 665–673 (2020)
7. González, C., et al.: Distance-based detection of out-of-distribution silent failures for Covid-19 lung lesion segmentation. Med. Image Anal. **82**, 102596 (2022)

8. González, C., Lemke, N., Sakas, G., Mukhopadhyay, A.: What is wrong with continual learning in medical image segmentation? arXiv:2010.11008 (2020)
9. González, C., Ranem, A., Othman, A., Mukhopadhyay, A.: Task-agnostic continual hippocampus segmentation for smooth population shifts. In: MICCAI Workshop on Domain Adaptation and Representation Transfer, pp. 108–118 (2022)
10. González, C., Ranem, A., Pinto dos Santos, D., Othman, A., Mukhopadhyay, A.: Lifelong nnU-Net: a framework for standardized medical continual learning. Nat. Sci. Rep. **13**(1), 9381 (2023)
11. Guo, C., Pleiss, G., Sun, Y., Weinberger, K.Q.: On calibration of modern neural networks. In: International Conference on Machine Learning (2017)
12. Hendrycks, D., Gimpel, K.: A baseline for detecting misclassified and out-of-distribution examples in neural networks. In: International Conference on Learning Representations (2017)
13. Hong, Y., Mundt, M., Park, S., Uh, Y., Byun, H.: Return of the normal distribution: flexible deep continual learning with variational auto-encoders. Neural Netw. **154**, 397–412 (2022)
14. Isensee, F., Jaeger, P.F., Kohl, S.A., Petersen, J., Maier-Hein, K.H.: nnU-Net: a self-configuring method for deep learning-based biomedical image segmentation. Nat. Methods **18**(2), 203–211 (2021)
15. Kendall, A., Badrinarayanan, V., Cipolla, R.: Bayesian SegNet: model uncertainty in deep convolutional encoder-decoder architectures for scene understanding. In: British Machine Vision Conference (2017)
16. Kingma, D.P., Ba, J.: Adam: a method for stochastic optimization. In: International Conference on Learning Representations (2014)
17. Kingma, D.P., Welling, M.: Auto-encoding variational Bayes. In: International Conference on Learning Representations (2014)
18. Kirkpatrick, J., et al.: Overcoming catastrophic forgetting in neural networks. Proc. Natl. Acad. Sci. **114**(13), 3521–3526 (2017)
19. Kulaga-Yoskovitz, J., et al.: Multi-contrast submillimetric 3 tesla hippocampal subfield segmentation protocol and dataset. Scientific data **2**(1), 1–9 (2015)
20. Lee, K., Lee, H., Lee, K., Shin, J.: Training confidence-calibrated classifiers for detecting out-of-distribution samples. In: International Conference on Learning Representations (ICLR) (2018)
21. Lennartz, J., Schultz, T.: Segmentation distortion: quantifying segmentation uncertainty under domain shift via the effects of anomalous activations. In: International Conference on Medical Image Computing and Computer-Assisted Intervention, pp. 316–325 (2023)
22. Liang, S., Li, Y., Srikant, R.: Enhancing the reliability of out-of-distribution image detection in neural networks. In: International Conference on Learning Representations (2018)
23. Liu, Q., Dou, Q., Yu, L., Heng, P.A.: MS-Net: multi-site network for improving prostate segmentation with heterogeneous MRI data. IEEE Trans. Med. Imaging **39**(9), 2713–2724 (2020)
24. McCloskey, M., Cohen, N.J.: Catastrophic interference in connectionist networks : the sequential learning problem. Psychol. Learn. Motiv. Adv. Res. Theory **24**(C), 109–165 (1989)
25. Mundt, M., Hong, Y., Pliushch, I., Ramesh, V.: A wholistic view of continual learning with deep neural networks: forgotten lessons and the bridge to active and open world learning. Neural Netw. **160**, 306–336 (2023)
26. Parisi, G.I., Kemker, R., Part, J.L., Kanan, C., Wermter, S.: Continual lifelong learning with neural networks: a review. Neural Netw. **113**, 54–71 (2019)

27. Ranem, A., González, C., Mukhopadhyay, A.: Continual hippocampus segmentation with transformers. In: Proceedings of the IEEE/CVF Conference on Computer Vision and Pattern Recognition, pp. 3711–3720 (2022)

28. Ranem, A., González, C., dos Santos, D.P., Bucher, A.M., Othman, A.E., Mukhopadhyay, A.: Continual atlas-based segmentation of prostate MRI. In: Proceedings of the IEEE/CVF Winter Conference on Applications of Computer Vision (2024)

29. Ratcliff, R.: Connectionist models of recognition memory: constraints imposed by learning and forgetting functions. Psychol. Rev. **97**(2), 285–308 (1990)

30. Robins, A.: Catastrophic forgetting, rehearsal and pseudorehearsal. Connect. Sci. **7**(2), 123–146 (1995)

31. Rolnick, D., Ahuja, A., Schwarz, J., Lillicrap, T.P., Wayne, G.: Experience replay for continual learning. In: Neural Information Processing Systems (NeurIPS) (2018)

32. Ronneberger, O., Fischer, P., Brox, T.: U-Net: convolutional networks for biomedical image segmentation. In: Navab, N., Hornegger, J., Wells, W.M., Frangi, A.F. (eds.) MICCAI 2015. LNCS, vol. 9351, pp. 234–241. Springer, Cham (2015). https://doi.org/10.1007/978-3-319-24574-4_28

33. Sahiner, B., Chen, W., Samala, R.K., Petrick, N.: Data drift in medical machine learning: implications and potential remedies. Br. J. Radiol. **96**, 20220878 (2023)

34. Shin, H., Lee, J.K., Kim, J., Kim, J.: Continual learning with deep generative replay. In: Advances in Neural Information Processing Systems, vol. 30 (2017)

35. The European Commission: Regulation (EU) 2016/679 of the European parliament and of the council (2016). https://eur-lex.europa.eu/eli/reg/2016/679/oj/deu

36. Wisse, L.E., Daugherty, A.M., Olsen, R.K., Berron, D., Carr, V.A., Stark, C.E., Amaral, R.S., Amunts, K., Augustinack, J.C., Bender, A.R., et al.: A harmonized segmentation protocol for hippocampal and parahippocampal subregions: Why do we need one and what are the key goals? Hippocampus **27**(1), 3–11 (2017)

37. Zenke, F., Poole, B., Ganguli, S.: Continual learning through synaptic intelligence. In: International Conference on Machine Learning, pp. 3987–3995 (2017)

Exploring Wearable Emotion Recognition with Transformer-Based Continual Learning

Federica Rizza(✉), Giovanni Bellitto(✉) ⓘD, Salvatore Calcagno(✉) ⓘD, and Simone Palazzo(✉) ⓘD

PeRCeiVe Lab, University of Catania, Catania, Italy
{federica.rizza,giovanni.bellitto,simone.palazzo}@unict.it,
salvatore.calcagno@phd.unict.it

Abstract. The rapid advancement of wearable technology has enabled continuous, real-time health monitoring through devices such as smartwatches and fitness trackers. These devices generate vast amounts of biometric data, including heart rate, galvanic skin response (GSR), and activity levels, which can be used for personalized healthcare applications such as emotion recognition and stress monitoring. However, the use of medical data introduces privacy challenges due to regulations like HIPAA and GDPR, necessitating innovative learning techniques that do not rely on large, centralized datasets. Continual learning (CL) offers a solution by enabling models to incrementally acquire knowledge over time, adapting to new data without forgetting previously learned information. This paper evaluates the effectiveness of CL techniques in the context of emotion recognition using GSR data from the DEAP dataset. Each subject is treated as a separate task, and a custom transformer-based PatchTST model is trained sequentially on each patient's data. Results show that the CL approach achieves performance levels comparable to traditional methods that train on all patients' data simultaneously. This demonstrates the potential of CL to maintain high accuracy while preserving patient data privacy, thereby supporting the development of adaptive, real-time personalized healthcare solutions.

Keywords: Emotion recognition · Continual learning · Personalized medicine

1 Introduction

The rapid advancement of wearable technology has brought about a new era of health monitoring, offering access to continuous, real-time biometric data that was previously unattainable [11]. Wearable devices, such as smartwatches and fitness trackers, are equipped with sensors that are capable of measuring a variety of physiological parameters, including heart rate, galvanic skin response (GSR), body temperature, and activity levels. The generation of vast amounts of data

F. Proietto Salanitri et al. (Eds.): PILM 2024/AIPAD 2024, LNCS 15197, pp. 86–101, 2025.
https://doi.org/10.1007/978-3-031-73483-0_8

by these devices offers valuable insights into various aspects of an individual's health and well-being. One of the principal advantages of wearable technology is its capacity to facilitate a multitude of applications that extend well beyond the scope of traditional health monitoring. The data collected from these devices can be employed in a variety of ways, including the assessment of human well-being, the recognition of emotions [42], the monitoring of stress levels [5], and the detection of anomalies in physical activity patterns [45]. The utilisation of these capabilities is facilitating a transformation in the approach to personalised healthcare and well-being. The field of personalized medicine, which aims to tailor healthcare interventions to the unique characteristics of each individual, can greatly benefit from the continuous data stream provided by wearable devices [49]. The data thus generated allows for the creation of highly detailed and individualised health profiles, which in turn enables healthcare providers to deliver more targeted and effective treatments. Moreover, the incorporation of wearable data with other personal health information increases the precision and personalisation of medical care.

Artificial intelligence (AI) models are of great importance in the analysis of the complex and voluminous data generated by wearable devices. These models, that can employ machine learning (ML) and deep learning techniques (DL), are highly proficient at identifying patterns and making predictions based on biometric signals. For example, AI algorithms can be trained to recognise emotional states, evaluate stress levels and provide real-time assessments of physical and mental health. These capabilities are vital for developing responsive and adaptive healthcare solutions that can proactively address the needs of individuals.

The application of traditional learning methods in AI, such as supervised learning, has yielded significant results in the processing and analysis of data collected from wearable device. These methods entail training a model on a substantial, labelled dataset, thereby enabling it to make accurate predictions or classifications based on new, previously unseen data. However, the nature of medical data introduces unique challenges that necessitate considerations beyond those of conventional applications. One of the primary concerns is patient privacy. Medical data is inherently sensitive, containing personal and health information that must be protected to comply with regulations such as the Health Insurance Portability and Accountability Act (HIPAA) in the United States and the General Data Protection Regulation (GDPR) in the European Union. These regulations impose stringent requirements on how personal health data can be collected, stored, and shared. As a result, it is often impractical or even illegal to retain and utilize large datasets of patient information for extended periods or across multiple studies.

To address these privacy and security concerns, AI models in medicine must be capable of operating effectively even in scenarios where access to data from previous patients is restricted. This paradigm shift calls for the development of innovative learning techniques that do not rely on the retention of large, centralized datasets. In this setting, continual learning (CL), also known as lifelong learning, becomes a key concept. CL models are designed to facilitate the incremental acquisition of knowledge over time, retaining previously acquired information while adapting their representation of information as new data becomes available [24]. This approach is particularly advantageous in the context of personalized medicine for several reasons:

- **Personalised Adaptation**: CL enables models to evolve in accordance with the individual, adapting to changes in their health status, lifestyle, and environment. This guarantees that the AI system remains pertinent and precise over time.
- **Domain Shift**: Patients have unique health profiles and their data can vary significantly [36]. CL can accommodate this diversity by learning from each individual's data separately while retaining generalizable knowledge that can benefit future tasks.
- **Resource Efficiency**: CL has the potential to reduce the necessity for retraining models from scratch with each new dataset. Instead, the model can incrementally update its knowledge base, thereby making the process more resource-efficient and scalable.
- **Improved Generalization**: By learning continuously from a wide array of data, continual learning models can improve their ability to generalize across different patients and scenarios, enhancing their overall robustness and effectiveness.

Furthermore, CL addresses two significant challenges inherent to medical data science: the limited size of medical datasets and the incremental availability of patient data. Medical datasets are frequently comprised primarily of a limited number of samples, due to the difficulties and costs associated with the collection of high-quality medical data. The scarcity of data can present a challenge for traditional AI models, which often require large datasets to achieve high accuracy. However, CL models are capable of learning effectively from smaller, sequentially available datasets by incrementally building on previously acquired knowledge. Moreover, in practical medical settings, it is frequently unfeasible to gain access to all patient data from the outset. Instead, data is received incrementally as patients undergo new examinations, treatments, and monitoring. CL models are well-suited to this context, as they can integrate new data into their learning process without requiring retraining from scratch. This capability ensures that the AI model remains up to date with the latest patient information, thereby enhancing its relevance and accuracy in making personalised healthcare decisions.

In this paper, we aim to evaluate the effectiveness of continual learning techniques within the medical context, specifically focusing on emotion recognition. Our objective is to emulate a real-world setting where patient conditions are continuously monitored through wearable devices, using Galvanic Skin Response (GSR) as the primary biomedical signal. GSR is a valuable indicator of emotional states, providing a direct measure of skin conductance that varies with sweating, which is influenced by emotional arousal.

To achieve this, we simulate a continual learning environment using the DEAP [22] dataset, a well-established dataset for emotion recognition. In our setup, each subject in the dataset is treated as a separate task. We train a model on each patient's data individually and sequentially, reporting the model's performance after it has processed data from all patients. This approach closely mimics a real-world scenario where new patient data becomes available over time, necessitating the model to adapt continuously without forgetting previously learned information. We employ a custom transformer-based PatchTST [32] model, which is particularly suitable for processing time series data like the signals provided by GSR sensors.

Our results demonstrate that training an emotion recognition model in a continual learning setting can achieve performance levels comparable to those attained by training with all patients' data simultaneously. This finding is significant, as it validates the potential of continual learning techniques to maintain high accuracy while adapting to new data incrementally. Our approach not only preserves the integrity of individual patient data in compliance with privacy regulations but also supports the development of personalized healthcare solutions that can evolve over time with the patient's condition.

2 Related Works

2.1 Emotion Recognition

The conceptual framework for understanding and categorizing emotional states is well-established. According to the study by Russell and Mehrabian, emotional states can be defined along three independent dimensions: valence (positive/pleasurable or negative/unpleasurable), arousal (engaged or not engaged), and dominance (degree of control that a person has over their affective states) [41]. Leveraging this theory, various machine learning (ML) techniques have been explored for emotion recognition and the assessment of specific emotions. For instance, Mitrut et al. [31] and others [9] have utilized different ML approaches to identify and classify emotional states, using the valence and arousal levels to detect stress or calm conditions in subjects [44].

Traditional ML techniques have been extensively used for emotion recognition. Kusumaningrum et al. [23] compared Support Vector Machine (SVM) and Random Forest methods for emotion classification using features extracted from EEG signals in the DEAP dataset. Similarly, another study [34] utilized a stacked autoencoder for feature extraction from EEG data from both the DEAP and SEED [55] datasets, followed by a three-level classification using a soft voting strategy among Decision Tree, KNN, and Random Forest algorithms.

Additional studies have employed Principal Component Analysis (PCA) for feature extraction, followed by SVM and KNN classifiers to analyze EEG data from the DEAP dataset [6]. Taneja et al. [46] integrated eye tracking information with EEG data from the SEED dataset [55] and evaluated emotions using SVM and KNN methods. In contrast, a study by authors in [35] employed KNN, Naive Bayes, and SVM classifiers on a custom database of songs to estimate the emotions induced in listeners.

Data from wearable devices have also been utilized for emotion recognition. For example, a study [14] used EEG, blood pressure, SPO2, ECG, and temperature data recorded from wearable devices and performed classification using a Reputation-Driven SVM.

Deep learning approaches have shown significant promise in emotion recognition tasks. A novel attention-based LSTM system proposed by [53] integrates data from a variety of sensors, including a smartphone's front camera, microphone, and touch panel, as well as a wristband measuring photoplethysmography, electrodermal activity, and infrared thermopile sensor, to accurately assess the user's emotional state. The system was evaluated through a user study with 45 participants.

In another study, Hassan et al. [19] proposed a Fine Gaussian SVM and Deep Belief Network (DBN) architecture for recognizing emotions using Electro-Dermal Activity (EDA), Photoplethysmogram (PPG), and Zygomaticus Electromyography (zEMG) sensor signals from the DEAP dataset. A Bimodal Deep AutoEncoder (BDAE) combining EEG and eye tracker signals from the DEAP and SEED datasets was adopted in [25].

A Residual LSTM was proposed by authors in [27] to learn the correlation between EEG and other physiological signals from the DEAP dataset. Another notable approach is the Bimodal-LSTM introduced in [47], which was evaluated on SEED and DEAP datasets using EEG features and eye movement features as input. Finally, Aiswaryadevi et al. [2] detected arousal levels using a Convolutional Neural Network (CNN) and Recurrent Neural Network (RNN) with multimodal data, including text, images, and acoustic signals.

2.2 Continual Learning

Continual Learning (CL) [15, 30, 33] addresses the problem of *catastrophic forgetting* in neural networks, wherein they tend to lose previously acquired knowledge when faced with shifts in input data distribution. To compensate for this problem, countless solutions have been proposed that introduce either adequate regularization terms [21, 54], specific architectural organization [29, 43] or the rehearsal of a small number of previously encountered data points [8, 38, 40].

Continual learning problems are generally classified into three broad categories: class-incremental continual learning, task-incremental continual learning, and domain-incremental continual learning. Class-incremental continual learning involves scenarios where new classes of data are introduced over time, and the model must learn to recognize these new classes without forgetting the previously learned classes. This is generally considered the most challenging scenario, since at inference time the model has no information on the task (i.e., which subset of classes) an input sample comes from. Task-incremental continual learning also assumes that the class distribution between tasks may differ; however, the model generally has knowledge of the specific task an input sample comes from, which allows for using task-specific heads or modules to manage the learning process. Domain-incremental continual learning deals with the scenario where the input domain changes over time, but the task and classes remain the same: the challenge in this setting is to adapt to the new domains without degrading performance on the previous ones, for a fixed classification problem.

To reach reasonable performance, most approaches tackling this challenging scenario adopt a replay strategy [37,40]. Some works focus on memory management: GSS [4] introduces a specific optimization of the basic rehearsal formula meant to store maximally informative samples in memory, while HAL [12] individuates synthetic replay data points that are maximally affected by forgetting. Other approaches propose tailored classification schemes: CoPE [16] uses class prototypes to ensure a gradual evolution of the shared latent space; ER-ACE [10] makes the cross-entropy loss asymmetric to minimize imbalance between current and past tasks. Finally, other works introduce a surrogate optimization objective: SCR [28] employs a supervised contrastive learning objective and OCM [18] leverages mutual information objectives: both aim at learning informative features that are less subject to forgetting. However, while current solutions help mitigating forgetting, their application to real-world settings proves difficult, as typical CL evaluations are conducted in accordance to unrealistic benchmarks [3,51].

2.3   Application of Continual Learning to Healthcare

In the field of medicine, a continually learning model could significantly aid clinicians with various tasks such as providing diagnoses and making management decisions. By incorporating new patient data along with the outcomes of previous tasks (such as actual diagnoses or treatment results), the model could apply its prior knowledge to the new data, refine its current tasks, incrementally learn to handle new tasks and even improve its own performance on past tasks, in light of newly-acquired information. At present, machine learning and deep learning applications in medical research are predominantly confined to supervised learning. To date, only a few automated algorithms have received approval from the US Food and Drug Administration (FDA) for specific uses, such as detecting diabetic retinopathy [1]. These algorithms are "locked" for safety reasons, meaning they cannot continue to learn or change after approval. However, models that employ continual (or "unlocked") learning could be more beneficial. These models would be able to learn from their mistakes and improve their performance over time as they are exposed to more data, much like human clinicians do. Investigating new methodological solutions for the application of continual learning in medicine is thus a crucial step in advancing healthcare technology.

3   Method

3.1   Problem Formulation

In our work, we tackle continual learning as a supervised regression problem, with target predictions corresponding to estimation of emotional variables, under the assumption that data is presented to the model as a sequence of tasks with non-i.i.d. data distributions; we also assume that knowledge of passing from one task to another is available, which is consistent with our mapping of tasks onto individual patients.

Formally, let $\mathcal{D} = \{\mathcal{D}_1, \ldots, \mathcal{D}_T\}$ be the sequence of task data distributions, from which it is possible to sample a data point $(\mathbf{x}, y) \sim \mathcal{D}_i$, where $\mathbf{x} \in \mathcal{X}$ is an observation with associated target $y \in \mathcal{Y}$. Our scenario thus corresponds to a domain-incremental learning problem, as $\mathcal{Y}$ is the same for all tasks.

Let $f : \mathcal{X} \rightarrow \mathcal{Y}$ be a regression model parameterized by $\boldsymbol{\theta}$; our objective is to train f on $\mathcal{D}$, organized as a sequence of T tasks $\{\tau_1, \ldots, \tau_T\}$. At each task τ_i, we train the model on input samples from the corresponding data distribution, i.e., $(\mathbf{x}, y) \sim \mathcal{D}_i$. In our setting, we take into consideration *rehearsal-based* continual learning methods only [8,39], as they are the ones that consistently outperform other variants. As a consequence, we assume that the regression model may also keep a limited *memory buffer* $\mathbf{M}$ of past samples, to reduce forgetting of features from previous tasks. The training phase at each task can be denoted as an update step of the model:

$$\langle f, \boldsymbol{\theta}_{i-1}, \mathbf{M}_{i-1} \rangle \xrightarrow{\mathcal{D}_i} \langle f, \boldsymbol{\theta}_i, \mathbf{M}_i \rangle \tag{1}$$

with $\boldsymbol{\theta}_i$ being the set of model parameters and $\mathbf{M}_i$ the memory buffer, at the end of task τ_i.

Our training objective is to optimize a standard mean square error loss over the entire sequence of tasks, computed on the model obtained after finishing the training procedure:

$$\arg\min_{\boldsymbol{\theta}_T} \sum_{i=1}^{T} \mathbb{E}_{(\mathbf{x},y) \sim \mathcal{D}_i} \left[\mathcal{L}\Big(f(\mathbf{x}; \boldsymbol{\theta}_T), y \Big) \right] \tag{2}$$

where $\mathcal{L}$ is the regression loss.

3.2   Model

Given that biosignals are recorded as time series, we have selected a transformer-based architecture [50] known as PatchTST [32] for our emotion recognition task. PatchTST, whose architecture is illustrated in Fig. 1, is specifically designed to handle prediction and classification on time series, making it an ideal choice for our task. The foundation of PatchTST lies in its approach to segmenting time series data: instead of working on the individual samples directly, the model divides the time series into smaller subseries-level units, or "patches", similarly to what is done in vision transformers. These patches are then projected to the target dimensionality and serve as input tokens to the transformer. This segmentation not only simplifies the data structure but also enhances the model's ability to capture local temporal patterns within each patch.

PatchTST is designed to operate on multivariate time series independently of channel dependencies, by treating each channel as a separate univariate time series. While this may be a limitation of the approach, as it disregards inter-relationships between channels, in our application we use Galvanic Skin Response (GSR) as the sole data source, making this shortcoming negligible.

The patching mechanism in PatchTST allows to preserve the local semantic information within each segment, and leads to a quadratic reduction in the computation and memory usage of attention maps, provided that the look-back window remains constant. This efficiency is particularly beneficial for real-time processing and deployment on resource-constrained wearable devices. Moreover, segmentation allows the model to attend to a longer historical record, enabling to capture long-term dependencies and trends in the biosignal data, which are often indicative of emotional states.

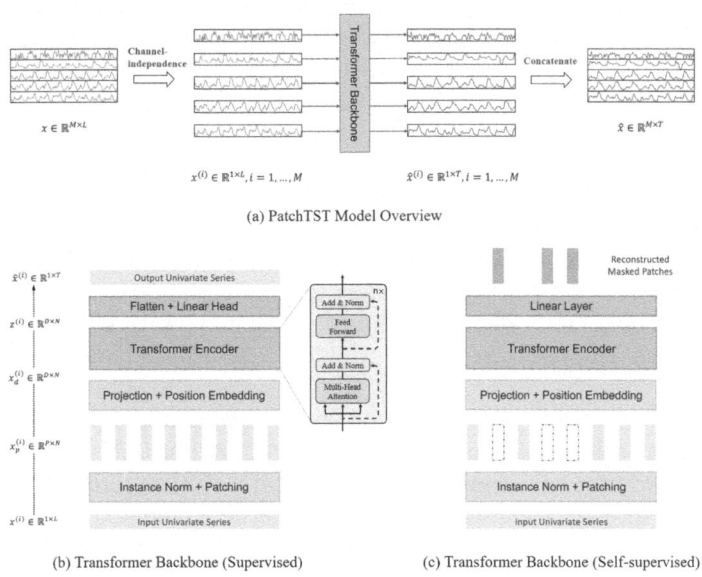

(a) PatchTST Model Overview

(b) Transformer Backbone (Supervised)   (c) Transformer Backbone (Self-supervised)

Fig. 1. PatchTST Architecture

4 Experimental Results

4.1 Dataset

We use the public dataset DEAP (Database for Emotion Analysis using Physiological Signals) [22], composed by physiological signals of 32 participants and frontal face video of the first 22 participants. All of these data were recorded as each subject watched 40 one-minute long music video. For each video, participants performed a self-assessment of their levels of valence, arousal, dominance and liking as a float value between 1 and 9, and familiarity as an integer between 1 and 5. The first two quantities are the primary focus of this study, as they are effective in discerning emotional states or stress levels [52]. During the procedure, electroencephalogram (EEG) was recorded as well as other thirteen peripheral physiological signals, like GSR, respiration amplitude, skin temperature, electrocardiogram, blood volume by plethysmograph, electromyograms of Zygomaticus and Trapezius muscles, and electrooculogram (EOG). As stated in [20], GSR value alteration is bound to emotional stimulation, and features extracted from this signal yeld to good results [17]; in view of the above we employed this signal in our analysis.

For our experiments we decided to evaluate the GSR of 5 subjects, with IDs 12 to 16, as our goal is to obtain good performance for each of them indipendently. Also, a split of the dataset, formed by the first 10 subject, was used for pretraining the backbone of the model in one of our scenarios. Each subject's dataset contains 40 time series, so we split each one into test, validation and test sets with a 65/10/25 proportion when we perform the standard supervised training. The validation set is not required during continual training, so a 75/25 ratio is employed for the division of each subject's dataset into a training and testing set.

4.2   Signal Pre-processing

The preprocessed data from DEAP have a sample rate of 128 Hz; in this work, we removed the first 3 s, corresponding to a pre-training baseline, and then we apply an order 8 Chebyshev type I filter to the data, which, at the end, were downsampled by a factor of 100, obtaining 77 sample per time-series. This preprocessing does not result in any alteration to the signal pattern in the time domain.

As we perform two-level classification on valence and arousal dimensions, the labels from the self-assessment are changed into "low" or 0 for values less than 5 and "high" or 1 for values greater then or equal to 5. The resulting signals are standardized through mean and standard deviation computed on the training set. At training time, data augmentation is carried out by extracting random crops from the signal, with crop size equal to 65 samples. At test time, a central crop is extracted, of the same size.

4.3   Metrics

We report our results using several key metrics: *final average accuracy* (FAA), which is calculated as the mean accuracy across all tasks after training on the final task; *final average forgetting* (FAF), quantifying the average decline in performance on previous tasks from their peak accuracy to their final accuracy [13]; *final backward transfer* (FBT), which, contrary to FAF, evaluates the increase in accuracy on previously learned tasks [26]; and *final forward transfer* (FFT), which assesses the improvement in the model's performance on tasks that have not yet been encountered [26].

4.4   Training Procedure

In our experiments, we utilize PyTorch as deep learning framework and employed the implementation of PatchTST from Huggingface. The experiments are conducted within the continual learning paradigm using the Mammoth framework

[7,8], which we extended to incorporate the DEAP dataset and the PatchTST backbone.

Given the limited availability of data and our decision not to use a model pretrained on other datasets, we opt for a simpler model configuration than the default one, which is more prone to overfitting; the parameters of the model we employ are detailed in Table 1. We explore two variants of this model: one that is trained from scratch and another that undergoes pretraining using masked self-supervision on data from the first 10 subjects.

Subsequently, the best variant is subjected to supervised training in a continual learning setting on data from five subjects, specifically subjects 12 to 16. In this setting, we treat each subject within the DEAP dataset as an individual task. The PatchTST model is trained sequentially on the data from each subject, using continual learning algorithms such as DER++ [8], ER [39] and FDR [48], to enable the incremental acquisition of new knowledge while mitigating the risk of catastrophic forgetting of previously learned information. This methodology adheres to the objective of preserving patient data privacy by eliminating the necessity for a large, centralized dataset, thus ensuring that the model can learn effectively from each subject's data in isolation.

Table 1. PatchTST parameters

Encoder layers	2
Attention heads	2
d_{model}	32
Context length	65
Patch length	5
Patch stride	5
Use CLS token	True
Num. classes	2

4.5   Results

Tables 2 and 3 provide an overview of the performance of the continual learning methods under analysis, in terms of valence and arousal accuracy; each method's performance is assessed with varying buffer sizes. As baselines, we use *SGD* (i.e., training sequentially on each task, with no mitigation of catastrophic forgetting) and *Joint* (i.e., training on the union of all task datasets). For the *joint* scenario, we evaluate the usage of a pretrained model through masked self-supervision on a left-out set of subjects, compared to training from scratch. Best performance is provided by the pretrained model: for this reason, and also in light of additional preliminary experiments, we also employ the pretrained model as an initialization point for continual learning methods.

It is possible to notice that the impact of catastrophic forgetting is significant: results achieved in the *SGD* setting are markedly worse than the *joint* baseline. The introduction of continual learning approaches helps mitigating the drop in performance, and in some cases even improves performance compared to the *joint* reference, which is a remarkable result. It should be noted that this is partly due to our domain-incremental setting, which is generally easier than the more common class-incremental setting, for which the methods under analysis were designed. In this context, we also experience significant forward and backward transfers in certain configurations. The effect of buffer size is noticeable, although in some cases more than others.

Overall, our preliminary analysis demonstrates a positive impact of the introduction of continual learning methods in the context of personalized learning.

Table 2. Valence accuracy in a continual learning setting.

Method	Buffer size	FAA	FFT	FBT	FAF
DER++	200	59.33	−0.28	10.0	7.50
	500	67.56	10.28	23.06	0.00
ER	200	65.56	2.78	14.72	2.78
	500	67.56	3.06	20.0	5.28
FDR	200	57.33	10.83	5.56	15.56
	500	59.11	8.61	4.44	18.61
SGD	-	52.56	8.33	10.0	12.78
Joint (from scratch)	-	53.06	-	-	-
Joint (pretrained)	-	61.22	-	-	-

Table 3. Arousal accuracy in a continual learning setting.

Method	Buffer size	FAA	FFT	FBT	FAF
DER++	200	54.67	20.56	−0.56	15.56
	500	55.89	20.56	−7.78	15.28
ER	200	56.89	35.28	−5.28	20.28
	500	63.11	23.06	2.50	5.00
FDR	200	57.33	30.28	0.56	15.0
	500	63.11	23.06	0.28	12.78
SGD	-	52.67	23.06	7.22	13.06
Joint (from scratch)	-	53.16	-	-	-
Joint (pretrained)	-	56.94	-	-	-

5   Conclusions

In this paper, we propose an analysis on the use of Continual Learning in the field of personalized medicine, highlighting its potential to address privacy concerns and the challenge of not having access to the entire dataset from the outset. Our study focused on emotion recognition using galvanic skin response data from the DEAP dataset, employing a custom transformer-based PatchTST model.

The results indicate that while catastrophic forgetting significantly impacts performance in a standard stochastic gradient descent setting, the application of continual learning techniques can mitigate this issue. In some instances, continual learning approaches even surpassed the performance of traditional joint training methods, demonstrating their efficacy in handling sequentially available data. This remarkable outcome underscores the feasibility of continual learning in maintaining high accuracy without the need for centralized datasets, thus preserving patient privacy.

Continual learning offers a promising methodology for developing adaptive personalized healthcare solutions. By enabling models to learn incrementally from new data while retaining previously acquired knowledge, continual learning supports the creation of privacy-preserving systems capable of delivering high-accuracy health monitoring and emotion recognition. Future research should explore the application of continual learning in more complex and diverse medical datasets, as well as investigate its potential in other areas of personalized medicine.

Acknowledgments. The work of Federica Rizza, who has contributed to the development of model architecture and experimental evaluations, has been supported by MUR in the framework of PNRR Mission 4, Component 2, Investment 1.1, PRIN, under project RESILIENT, CUP E53D23016360001. Salvatore Calcagno and Giovanni Bellitto acknowledges financial support from PNRR MUR project PE0000013-FAIR.The work of Simone Palazzo, who has contributed to the overall idea, experimental design and paper writing, has been supported by MUR in the framework of PNRR Mission 4, Component 2, Investment 1.1, PRIN, under project RESILIENT, CUP E53D23016360001.

Disclosure of Interests. The authors have no competing interests to declare that are relevant to the content of this article.

References

1. Abràmoff, M.D., Lavin, P.T., Birch, M., Shah, N., Folk, J.C.: Pivotal trial of an autonomous AI-based diagnostic system for detection of diabetic retinopathy in primary care offices. NPJ Digit. Med. **1**(1), 39 (2018)
2. Aiswaryadevi, V.J., et al.: Smart IoT multimodal emotion recognition system using deep learning networks. In: Manoharan, K.G., Nehru, J.A., Balasubramanian, S. (eds.) Artificial Intelligence and IoT. Studies in Big Data, vol. 85, pp. 3–19. Springer, Singapore (2021). https://doi.org/10.1007/978-981-33-6400-4_1

3. Aljundi, R., Kelchtermans, K., Tuytelaars, T.: Task-free continual learning. In: Proceedings of the IEEE/CVF Conference on Computer Vision and Pattern Recognition, pp. 11254–11263 (2019)

4. Aljundi, R., Lin, M., Goujaud, B., Bengio, Y.: Gradient based sample selection for online continual learning. In: Advances in Neural Information Processing Systems (2019)

5. Bakker, J., Pechenizkiy, M., Sidorova, N.: What's your current stress level? detection of stress patterns from GSR sensor data, pp. 573–580 (2011). https://doi.org/10.1109/ICDMW.2011.178

6. . Bazgir, O., Mohammadi, Z., Habibi, S.A.H.: Emotion recognition with machine learning using EEG signals. In: 2018 25th National and 3rd International Iranian Conference on Biomedical Engineering (ICBME), pp. 1–5 (2018). https://doi.org/10.1109/ICBME.2018.8703559

7. Boschini, M., Bonicelli, L., Buzzega, P., Porrello, A., Calderara, S.: Class-incremental continual learning into the extended der-verse. IEEE Trans. Pattern Anal. Mach. Intell. **45**, 5497–5512 (2022)

8. Buzzega, P., Boschini, M., Porrello, A., Abati, D., Calderara, S.: Dark experience for general continual learning: a strong, simple baseline. In: Larochelle, H., Ranzato, M., Hadsell, R., Balcan, M.F., Lin, H. (eds.) Advances in Neural Information Processing Systems, vol. 33, pp. 15920–15930. Curran Associates, Inc. (2020)

9. Bălan, O., Moise, G., Moldoveanu, A., Leordeanu, M., Moldoveanu, F.: Fear level classification based on emotional dimensions and machine learning techniques. Sensors **19**(7) (2019). https://doi.org/10.3390/s19071738

10. Caccia, L., Aljundi, R., Asadi, N., Tuytelaars, T., Pineau, J., Belilovsky, E.: New insights on reducing abrupt representation change in online continual learning. In: International Conference on Learning Representations Workshop (2022)

11. Chan, M., Estève, D., Fourniols, J.Y., Escriba, C., Campo, E.: Smart wearable systems: Current status and future challenges. Artif. Intell. Med. **56**(3), 137–156 (2012). https://doi.org/10.1016/j.artmed.2012.09.003, https://www.sciencedirect.com/science/article/pii/S0933365712001182

12. Chaudhry, A., Gordo, A., Dokania, P., Torr, P., Lopez-Paz, D.: Using hindsight to anchor past knowledge in continual learning. In: Proceedings of the AAAI Conference on Artificial Intelligence (2021)

13. Chaudhry, A., Ranzato, M., Rohrbach, M., Elhoseiny, M.: Efficient lifelong learning with A-GEM. arXiv preprint arXiv:1812.00420 (2018)

14. Dai, Y., Wang, X., Li, X., Zhang, P.: Reputation-driven multimodal emotion recognition in wearable biosensor network. In: 2015 IEEE International Instrumentation and Measurement Technology Conference (I2MTC) Proceedings, pp. 1747–1752 (2015). https://doi.org/10.1109/I2MTC.2015.7151544

15. De Lange, M., et al.: A continual learning survey: defying forgetting in classification tasks. IEEE Trans. Pattern Anal. Mach. Intell. **44**, 3366–3385 (2021)

16. De Lange, M., Tuytelaars, T.: Continual prototype evolution: learning online from non-stationary data streams. In: IEEE International Conference on Computer Vision (2021)

17. Goshvarpour, A., Goshvarpour, A.: The potential of photoplethysmogram and galvanic skin response in emotion recognition using nonlinear features. Australas. Phys. Eng. Sci. Med. **43** (2019). https://doi.org/10.1007/s13246-019-00825-7

18. Guo, Y., Liu, B., Zhao, D.: Online continual learning through mutual information maximization. In: International Conference on Machine Learning (2022)

19. Hassan, M.M., Alam, M.G.R., Uddin, M.Z., Huda, S., Almogren, A., Fortino, G.: Human emotion recognition using deep belief network architecture. Inform. Fusion **51**, 10–18 (2019). https://doi.org/10.1016/j.inffus.2018.10.009, https://www.sciencedirect.com/science/article/pii/S1566253518301301
20. Kipli, K., et al.: Evaluation of Galvanic Skin Response (GSR) signals features for emotion recognition, pp. 260–274 (2023). https://doi.org/10.1007/978-3-031-24801-6_19
21. Kirkpatrick, J., et al.: Overcoming catastrophic forgetting in neural networks. Proc. Natl. Acad. Sci. **114**, 3521–3526 (2017)
22. Koelstra, S., et al.: DEAP: a database for emotion analysis using physiological signals. IEEE Trans. Affect. Comput. **3**, 18–31 (2011). https://doi.org/10.1109/T-AFFC.2011.15
23. Kusumaningrum, T.D., Faqih, A., Kusumoputro, B.: Emotion recognition based on DEAP database using EEG time-frequency features and machine learning methods. Journal of Physics: Conference Series **1501**(1), 012020 (2020). https://doi.org/10.1088/1742-6596/1501/1/012020, https://dx.doi.org/10.1088/1742-6596/1501/1/012020
24. Lesort, T., Lomonaco, V., Stoian, A., Maltoni, D., Filliat, D., Díaz-Rodríguez, N.: Continual learning for robotics: definition, framework, learning strategies, opportunities and challenges. Inform. Fusion **58**, 52–68 (2020). https://doi.org/10.1016/j.inffus.2019.12.004, https://www.sciencedirect.com/science/article/pii/S1566253519307377
25. Liu, W., Zheng, W.L., Lu, B.L.: Emotion recognition using multimodal deep learning. In: Hirose, A., Ozawa, S., Doya, K., Ikeda, K., Lee, M., Liu, D. (eds.) Neural Information Processing, pp. 521–529. Springer International Publishing, Cham (2016). https://doi.org/10.1007/978-3-319-46672-9_58
26. Lopez-Paz, D., Ranzato, M.: Gradient episodic memory for continual learning. In: Advances in Neural Information Processing Systems (2017)
27. Ma, J., Tang, H., Zheng, W.L., Lu, B.L.: Emotion recognition using multimodal residual LSTM network. In: Proceedings of the 27th ACM International Conference on Multimedia, p. 176–183. MM 2019, Association for Computing Machinery, New York, NY, USA (2019). https://doi.org/10.1145/3343031.3350871
28. Mai, Z., Li, R., Kim, H., Sanner, S.: Supervised contrastive replay: revisiting the nearest class mean classifier in online class-incremental continual learning. In: IEEE International Conference on Computer Vision and Pattern Recognition Workshops (2021)
29. Mallya, A., Lazebnik, S.: PackNet: adding multiple tasks to a single network by iterative pruning. In: Proceedings of the IEEE Conference on Computer Vision and Pattern Recognition (2018)
30. McCloskey, M., Cohen, N.J.: Catastrophic interference in connectionist networks: the sequential learning problem. Psychology of learning and motivation (1989)
31. Mitrut, O., Moise, G., Petrescu, L., Moldoveanu, A., Leordeanu, M., Moldoveanu, F.: Emotion classification based on biophysical signals and machine learning techniques. Symmetry **12**, 21 (2019). https://doi.org/10.3390/sym12010021
32. Nie, Y., Nguyen, N.H., Sinthong, P., Kalagnanam, J.: A time series is worth 64 words: long-term forecasting with transformers. In: International Conference on Learning Representations (2023)
33. Parisi, G.I., Kemker, R., Part, J.L., Kanan, C., Wermter, S.: Continual lifelong learning with neural networks: a review. Neural Netw. **113**, 54–71 (2019)

34. Qing, C., Qiao, R., Xu, X., Cheng, Y.: Interpretable emotion recognition using EEG signals. IEEE Access **7**, 94160–94170 (2019). https://doi.org/10.1109/ACCESS.2019.2928691
35. Quilingking Tomas, J., S. Jamilla, R.A., S. Lopo, K., E. Camba, C.: Multimodal emotion detection model implementing late fusion of audio and lyrics in Filipino music. In: 2020 the 3rd International Conference on Computing and Big Data, pp. 78–84. ICCBD 2020, Association for Computing Machinery, New York, NY, USA (2020). https://doi.org/10.1145/3418688.3418702
36. Quionero-Candela, J., Sugiyama, M., Schwaighofer, A., Lawrence, N.D.: Dataset shift in machine learning (2009). https://api.semanticscholar.org/CorpusID:61294087
37. Ratcliff, R.: Connectionist models of recognition memory: constraints imposed by learning and forgetting functions. Psychol. Rev. **97**, 285 (1990)
38. Rebuffi, S.A., Kolesnikov, A., Sperl, G., Lampert, C.H.: iCaRL: incremental classifier and representation learning. In: Proceedings of the IEEE Conference on Computer Vision and Pattern Recognition (2017)
39. Riemer, M., et al.: Learning to learn without forgetting by maximizing transfer and minimizing interference. In: ICLR (2019)
40. Robins, A.: Catastrophic forgetting, rehearsal and pseudorehearsal. Connection Sci. **7**, 123–146 (1995)
41. Russell, J.A., Mehrabian, A.: Evidence for a three-factor theory of emotions. J. Res. Pers. **11**(3), 273–294 (1977). https://doi.org/10.1016/0092-6566(77)90037-X, https://www.sciencedirect.com/science/article/pii/009265667790037X
42. Saganowski, S., et al.: Emotion recognition using wearables: a systematic literature review - work-in-progress. In: 2020 IEEE International Conference on Pervasive Computing and Communications Workshops (PerCom Workshops), pp. 1–6 (2020). https://doi.org/10.1109/PerComWorkshops48775.2020.9156096
43. Schwarz, J., et al.: Progress & compress: a scalable framework for continual learning. In: International Conference on Machine Learning (2018)
44. Shon, D., Im, K., Park, J.H., Lim, D.S., Jang, B., Kim, J.M.: Emotional stress state detection using genetic algorithm-based feature selection on EEG signals. Int. J. Environ. Res. Public Health **15**(11) (2018). https://doi.org/10.3390/ijerph15112461, https://www.mdpi.com/1660-4601/15/11/2461
45. Sunny, J., et al.: Anomaly detection framework for wearables data: a perspective review on data concepts, data analysis algorithms and prospects. Sensors **22**, 756 (2022). https://doi.org/10.3390/s22030756
46. Taneja, R., Singh, J., Gill, R.: Multimodal emotion recognition system using machine learning and psychological signals: a review. In: Sharma, T.K., Ahn, C.W., Verma, O.P., Panigrahi, B.K. (eds.) Soft Computing: Theories and Applications. Advances in Intelligent Systems and Computing, vol. 1380, pp. 657–666. Springer, Singapore (2022). https://doi.org/10.1007/978-981-16-1740-9_54
47. Tang, H., Liu, W., Zheng, W.L., Lu, B.L.: Multimodal emotion recognition using deep neural networks. In: Liu, D., Xie, S., Li, Y., Zhao, D., El-Alfy, E.S.M. (eds.) Neural Information Processing, pp. 811–819. Springer International Publishing, Cham (2017). https://doi.org/10.1007/978-3-319-70093-9_86
48. Titsias, M.K., Schwarz, J., Matthews, A.G.d.G., Pascanu, R., Teh, Y.W.: Functional regularisation for continual learning with gaussian processes. arXiv preprint arXiv:1901.11356 (2019)

49. Tyler, J., Choi, S.W., Tewari, M.: Real-time, personalized medicine through wearable sensors and dynamic predictive modeling: a new paradigm for clinical medicine. Curr. Opin. Syst. Biol. **20**, 17–25 (2020). https://doi.org/10.1016/j.coisb.2020.07.001, https://www.sciencedirect.com/science/article/pii/S2452310020300068
50. Vaswani, A., et al.: Attention is all you need. In: Guyon, I., et al. (eds.) Advances in Neural Information Processing Systems, vol. 30. Curran Associates, Inc. (2017)
51. van de Ven, G.M., Tuytelaars, T., Tolias, A.S.: Three types of incremental learning. Nat. Mach. Intell. **4**, 1185–1197 (2022)
52. Whissell, C.: Using the revised dictionary of affect in language to quantify the emotional undertones of samples of natural language. Psychol. Rep. **105**(2), 509–521 (2009). https://doi.org/10.2466/PR0.105.2.509-521, pMID: 19928612
53. Yang, K., et al.: Behavioral and physiological signals-based deep multimodal approach for mobile emotion recognition. IEEE Trans. Affect. Comput. **14**(2), 1082–1097 (2023). https://doi.org/10.1109/TAFFC.2021.3100868
54. Zenke, F., Poole, B., Ganguli, S.: Continual learning through synaptic intelligence. In: International Conference on Machine Learning (2017)
55. Zheng, W.L., Lu, B.L.: Investigating critical frequency bands and channels for EEG-based emotion recognition with deep neural networks. IEEE Trans. Auton. Ment. Dev. **7**(3), 162–175 (2015). https://doi.org/10.1109/TAMD.2015.2431497

Author Index

F. Proietto Salanitri et al. (Eds.): PILM 2024/AIPAD 2024, LNCS 15197, pp. 103–104, 2025.
https://doi.org/10.1007/978-3-031-73483-0